Neuroscience and Philosophy

Brain, Mind, and Language

MAXWELL BENNETT, DANIEL DENNETT,

PETER HACKER, JOHN SEARLE

With an Introduction and Conclusion by Daniel Robinson

Columbia University Press   New York

Columbia University Press
Publishers Since 1893
New York, Chichester, West Sussex
Excerpts from *Philosophical Foundations of Neuroscience*
by M. R. Bennett and P. M. S. Hacker
copyright © 2003 Maxwell R. Bennett and Peter M. S. Hacker
"Philosophy as Naive Anthropology: Comment on Bennett and Hacker"
by Daniel Dennett copyright © 2007 Daniel Dennett
"Putting Consciousness Back in the Brain: Reply to Bennett and Hacker,
Philosophical Foundations of Neuroscience" by John Searle
copyright © 2007 John Searle
"The Conceptual Presuppositions of Cognitive Neuroscience:
A Reply to Critics" by Maxwell R. Bennett and Peter Hacker
copyright © 2007 Maxwell R. Bennett and Peter M. S. Hacker
"Epilogue" by Maxwell R. Bennett
copyright © 2007 Maxwell R. Bennett
"Introduction" and "Still Looking: Science and Philosophy
in Pursuit of Prince Reason" by Daniel Robinson
copyright © 2007 Columbia University Press
All rights reserved

Library of Congress Cataloging-in-Publication Data

Neuroscience and philosophy : brain, mind, and language /
Maxwell Bennett ... [et al.].
p. ; cm.
Includes bibliographical references.
ISBN-13: 978–0–231–14044–7 (cloth ; alk. paper)
978-0-231-14045-4 (pbk. : alk. paper)
1. Cognitive neuroscience—Philosophy.
2. Bennett, M. R. Philosophical foundations of neuroscience.
I. Bennett, M. R.
[DNLM: 1. Cognitive Science. 2. Neuropsychology.
3. Philosophy. WL 103.5 N4948 2007]
QP360.5.N4975 2007
612.8'2—dc22   2006036008

Casebound editions of Columbia University Press books
are printed on permanent and durable acid-free paper.

Printed in the United States of America

Neuroscience and Philosophy

CONTENTS

REPLY TO THE REBUTTALS

INTRODUCTION

Philosophical Foundations of Neuroscience, by Max Bennett and Peter Hacker, was published by Blackwell in 2003. It attracted attention straightaway because it was the first systematic evaluation of the conceptual foundations of neuroscience, as these foundations had been laid by scientists and philosophers. What added to the attraction of the work were two appendixes devoted specifically and critically to the influential writings of John Searle and Daniel Dennett. Max Bennett, an accomplished neuroscientist, correctly identified Searle and Dennett as the philosophers most widely read within the neuroscience community and was eager to make clear to readers why he and Hacker disagreed with their views.

In the fall of 2004 Bennett and Hacker were invited by the program committee of the American Philosophical Association to participate in an "Authors and Critics" session at the 2005 meeting of the association in New York. The choice of critics could not have been better: Daniel Dennett and John Searle had agreed to write replies to the criticisms levied against their work by Bennett and Hacker. The contents of this present volume are based on that three-hour APA session. The session was chaired by Owen Flanagan and was marked by an unusually animated exchange among the participants. Dennett and Searle had provided written versions of their rebuttals prior to the session, to which Bennett and Hacker then replied.

Fully aware of the importance of the philosophical issues, Wendy Lochner, the philosophy editor for Columbia University

Press, urged the participants to consider having the proceedings published in book form. In the ordinary course of events, the written version of a spirited colloquium generally imposes a rather gray cast over what was originally colorful and affecting. The imagination of the reader is summoned to the task of re-creating a real event out of the shards and apparatus of published prose. I think it is fair to say that this customary limitation is not suffered by the present volume. Reader will recognize in these essays and exchanges the motivating power of intellectual passion. The participants are serious about their subject. Their notable contributions over a course of decades give them the right to be taken seriously. Moreover, the stakes are uncommonly high. After all, the project of cognitive neuroscience is nothing less than the incorporation of what we are pleased to call human nature into the framework of science itself. Dennett and Searle, with a confidence that may appear as eagerness, are inclined to believe that the process of incorporation is well on its way. Bennett and Hacker, with a cautiousness that may appear as skepticism, raise the possibility that the project itself is based on a mistake.

I was honored to be asked to write a closing chapter for the intended volume. Such settled views as I might have on this subject are summarized in that chapter as I weigh the agile thrusts and parries by the central figures in the debate. Readers will note, I hope with compassion, that very little is settled in my own mind. I recognize the definite commitment Searle and Dennett have to providing a workable and credible model of just how our mental life is realized by events under the skin. Norbert Wiener, one of the truly wise men of science, noted that the best material model of a cat is a cat—preferably the same cat. Nonetheless, without models—even those laced with anthropomorphic seasoning—the very clutter of the real world must thwart scientific progress in any field. There is no calculus or equation establishing the boundaries within which the imagination of the model builder must confine itself.

In the end, matters of this sort rise to the level of aesthetics. By this I do not suggest that there is less room for analytical rigor; philosophical analysis at its best is an aesthetic undertaking. This surely is what attracts the physicist and the mathematician to what is "elegant." Is it not aesthetics that establishes Occam's razor as the tool of choice for refinement, measure, proportionality, coherence? In just these respects, I am sure readers will find in the Bennett-Hacker critique—chiefly in Peter Hacker's philosophically rich and informed critique—not a tilt toward skepticism but a careful and, yes, *elegant* application of the better tools philosophers have fashioned.

This much said, it is important to go further and to acknowledge that our actually *lived* life is unlikely to disclose its full, shifting, often fickle and wondrously interior reality either to the truth table, the Turing machine or the anatomical blowpipe. It should never come as a surprise that the philosopher who often gave us the first words on an important subject may well also have the last words to say on it. I refer, of course, to Aristotle. We ought to seek precision in those things that admit of it. We are to choose tools suitable to the task at hand. In the end, our explanations must make intelligible contact with that which we seek to explain. The demographer who tells us with commendable accuracy that the average family contains 2.53 members feels no obligation to remind us that there is no 0.53 person. Such data do not presume to describe the nature of the items counted; their result is just that number. The point, of course, is that scientific precision or, for that matter, arithmetic precision, may tell us next to nothing about just what has been assayed with such precision. Here as elsewhere, the ruling maxim is *caveat emptor*.

Readers will approach this discourse with proper interest— even a hint of vanity—for it is about *them!* They bring their own aesthetic standards to bear on material of this kind. It is finally they who will decide whether the accounts offered make intelligible contact with what really matters. But a good jury is no better than the evidence at hand, guided in their deliberation by sound rules of evidence. Patient reader! Worthy juror! Here is some of the (cognitive neuroscience) evidence and an exceptionally clear presentation of the rules that might be applied to the weighing of it. No need to hurry with a verdict . . .

Daniel N. Robinson

NEUROSCIENCE AND PHILOSOPHY

THE ARGUMENT

PHILOSOPHICAL FOUNDATIONS
OF NEUROSCIENCE

The Introduction

M. R. BENNETT AND P.M.S. HACKER

Philosophical Foundations of Neuroscience presents the fruits of a co-operative project between a neuroscientist and a philosopher. It is concerned with the conceptual foundations of *cognitive* neuro-science—foundations constituted by the structural relationships among the psychological concepts involved in investigations into the neural underpinnings of human cognitive, affective and volitional capacities. Investigating logical relations among concepts is a philosophical task. Guiding that investigation down pathways that will illuminate brain research is a neuroscientific one. Hence our joint venture.

If we are to understand the neural structures and dynamics that make perception, thought, memory, emotion and intention-al behaviour possible, clarity about these concepts and categories is essential. Both authors, coming to this investigation from very different directions, found themselves puzzled by, and sometimes uneasy with, the use of psychological concepts in contemporary neuroscience. The puzzlement was often over what might be meant by a given neuroscientist's claims concerning the brain and the mind, or over why a neuroscientist thought that the experiments he had undertaken illuminated the psychological capacity being studied, or over the conceptual presuppositions of the questions asked. The unease was produced by a suspicion that in some cases concepts were misconstrued, or were mis-applied, or were stretched beyond their defining conditions of application. And the more we probed, the more convinced we became that despite the impressive advances in cognitive neuro-science, not all was well with the general theorizing.

Empirical questions about the nervous system are the prov-ince of neuroscience. It is its business to establish matters of fact concerning neural structures and operations. It is the task of *cognitive* neuroscience to explain the neural conditions that make perceptual, cognitive, cogitative, affective and volitional functions possible. Such explanatory theories are confirmed or infirmed by experimental investigations. By contrast, conceptual questions (concerning, for example, the concepts of mind or memory, thought or imagination), the description of the logi-cal relations between concepts (such as between the concepts of perception and sensation, or the concepts of consciousness and self-consciousness), and the examination of the structural relationships between distinct conceptual fields (such as between the psychological and the neural, or the mental and the behav-ioural) are the proper province of philosophy.

Conceptual questions antecede matters of truth and falsehood. They are questions concerning our *forms of representation*, not questions concerning the truth or falsehood of empirical state-ments. These forms are presupposed by true (*and* false) scientific statements and by correct (*and* incorrect) scientific theories. They determine not what is empirically true or false, but rather what does and what does not make sense. Hence conceptual questions are not amenable to scientific investigation and experimentation or to scientific theorizing. For the concepts and conceptual relationships in question are *presupposed* by any such investigations and theorizings. Our concern here is not with trade union demarcation lines, but with distinctions between logically different kinds of intellectual inquiry.[1]

Distinguishing conceptual questions from empirical ones is of the first importance. When a conceptual question is confused with a scientific one, it is bound to appear singularly refractory. It seems in such cases as if science should be able to discover the truth of the matter under investigation by theory and experiment—yet it persistently fails to do so. That is not surprising, since conceptual questions are no more amenable to empirical methods of investigation than problems in pure mathematics are solvable by the methods of physics. Furthermore, when empirical problems are addressed without adequate conceptual clarity, misconceived questions are bound to be raised, and misdirected research is likely to ensue. For any unclarity regarding the relevant concepts will be reflected in corresponding unclarity in the questions and hence in the design of experiments intended to answer them. And any incoherence in the grasp of the relevant conceptual structure is likely to be manifest in incoherences in the interpretation of the results of experiments.

Cognitive neuroscience operates across the boundary between two fields, neurophysiology and psychology, the respective concepts of which are categorially dissimilar. The logical or conceptual relations between the physiological and the psychological are problematic. Numerous psychological concepts and categories of concepts are difficult to bring into sharp focus. The relations between the mind and the brain, and between the psychological and the behavioural are bewildering. Puzzlement concerning these concepts and their articulations, and concerning these apparent 'domains' and their relations, has characterized neurophysiology since its inception.[2] In spite of the great advances in neuroscience at the beginning of the twentieth century at the hands of Charles Sherrington, the battery of conceptual questions popularly known as the mind-body or mind-brain problem remained as intractable as ever—as is evident in the flawed Cartesian views embraced by Sherrington and by such of his colleagues and protegés as Edgar Adrian, John Eccles and Wilder Penfield. Brilliant though their work unquestionably was, deep conceptual confusions remained.[3] Whether the current generation of neuroscientists has successfully overcome the conceptual confusions of earlier generations, or whether it has merely replaced one conceptual entanglement by others is the subject of our investigation in this book.

One such tangle is evident in the persistent ascription of psychological attributes to the brain. For while Sherrington and his protegés ascribed psychological attributes to the mind (conceived as a peculiar, perhaps immaterial, substance distinct from the brain), contemporary neuroscientists tend to ascribe the same range of psychological attributes to the brain (commonly, although not uniformly, conceived to be identical with the mind). But the mind, we argue[4], is neither a substance distinct from the brain nor a substance identical with the brain. And we demonstrate that ascription of psychological attributes to the brain is incoherent.[5] Human beings possess a wide range of psychological powers, which are exercised in the circumstances of life, when we perceive, think and reason, feel emotions, want things, form plans and make decisions. The possession and exercise of such powers define us as the kinds of animals we are. We may enquire into the neural conditions and concomitants for their possession and exercise. This is the task of neuroscience, which is discovering more and more about them. But its discoveries in no way affect the conceptual truth that these powers and their exercise in perception, thought and feeling *are attributes of human beings*, not of their parts—in particular, *not of their brains*. A human being is a psychophysical unity, an animal that can perceive, act intentionally, reason, and feel emotions, a language-using animal that is not merely conscious, but also self-conscious—not a brain embedded in the skull of a body. Sherrington, Eccles and Penfield conceived of human beings as animals in whom the mind, which they thought of as the bearer of psychological attributes, is in liaison with the brain. It is no advance over that misconception to suppose that the brain is a bearer of psychological attributes.

Talk of the brain's perceiving, thinking, guessing or believing, or of one hemisphere of the brain's knowing things of which the other hemisphere is ignorant, is widespread among contemporary neuroscientists. This is sometimes defended as being no more than a trivial *façon de parler*. But that is quite mistaken. For the characteristic form of explanation in contemporary cognitive neuroscience consists in ascribing psychological attributes to the brain and its parts *in order to explain* the possession of psychological attributes and the exercise (and deficiencies in the exercise) of cognitive powers by human beings.

The ascription of psychological, in particular cognitive and cogitative, attributes to the brain is, we show, also a source of much further confusion. Neuroscience can investigate the neural conditions and concomitants of the acquisition, possession and exercise of sentient powers by animals. It can discover the neural preconditions for the possibility of the exercise of distinctively human powers of thought and reasoning, of articulate memory and imagination, of emotion and volition. This it can do by patient inductive correlation between neural phenomena and the possession and exercise of psychological powers, and between neural damage and deficiencies in normal mental functions. What it *cannot* do is *replace* the wide range of ordinary psychological explanations of human activities in terms of reasons, intentions, purposes, goals, values, rules and conventions by neurological explanations.[6] And it *cannot* explain how an animal perceives or thinks by reference to the brain's, or some part of the brain's, perceiving or thinking. For it makes no sense to ascribe such psychological attributes to anything less than the animal as a whole. It is the animal that perceives, not parts of its brain, and it is human beings who think and reason, not their brains. The brain and its activities *make it possible* for *us*—not for *it*—to perceive and think, to feel emotions, and to form and pursue projects.

While the initial response of many neuroscientists to the accusation of conceptual confusion is to claim that the ascription of psychological predicates to the brain is a mere *façon de parler*, their reaction to the demonstrable fact that their explanatory theories *non-trivially* ascribe psychological powers to the brain is sometimes to suggest that this error is unavoidable due to the deficiencies of language. We confront this misconception[7] [and] show that the great discoveries of neuroscience *do not require* this misconceived form of explanation—that what has been discovered can readily be described and explained in our existing language. We demonstrate this by reference to the much discussed phenomena resultant upon commissurotomy, described (or, we suggest, misdescribed) by Sperry, Gazzaniga and others.[8]

In Part II of *Philosophical Foundations of Nuroscience* (henceforth PFN) we investigate the use of concepts of perception, memory, mental imagery, emotion and volition in current neuroscientific theorizing. From case to case we show that conceptual unclarity, failure to give adequate attention to the relevant conceptual structures, has often been the source of theoretical error and the grounds for misguided inferences. It is an error, a conceptual error, to suppose that perception is a matter of apprehending an *image* in the mind (Crick, Damasio, Edelman), or the production of a *hypothesis* (Helmholtz, Gregory), or the generation of a *3-D model description* (Marr). It is confused, a conceptual confusion, to formulate the binding problem as the problem of combining data of shape, colour and motion to form *the image* of the object perceived (Crick, Kandel, Wurtz). It is wrong, conceptually wrong, to suppose that memory is always of the past, or to think that memories can be *stored* in the brain in the form of the strength of synaptic connections (Kandel, Squire, Bennett). And it is mistaken, conceptually mistaken, to suppose that investigating thirst, hunger and lust is an investigation into the emotions (Roles) or to think that the function of the emotions is to inform us of our visceral and musculoskeletal state (Damasio).

The initial reaction to such critical remarks may well be in-dignation and incredulity. How can a flourishing science be fundamentally in error? How could there be unavoidable conceptual confusion in a well-established science? Surely, if there are problematic concepts, they can easily be replaced by others that are unproblematic and that serve the same explanatory purposes.—Such responses betoken a poor understanding of the relation between form of representation and facts represented, and a misunderstanding of the nature of conceptual error. They also betray ignorance of the history of science in general and of neuroscience in particular.

Science is no more immune to conceptual error and confusion than any other form of intellectual endeavour. The history of science is littered with the debris of theories that were not simply factually mistaken, but conceptually awry. Stahl's theory of combustion, for example, was *conceptually* flawed in ascribing, in certain circumstances, negative weight to phlogiston—an idea that made no sense within its framework of Newtonian physics. Einstein's famous criticisms of the theory of electromagnetic aether (the alleged medium by which light was thought to be propagated) were not directed only at the results of the Michelson-Morley experiment which had failed to detect any effect of absolute motion, but also at a conceptual confusion concerning relative motion involved in the role ascribed to aether in the explanation of electromagnetic induction. Neuroscience has been no exception—as we show in our historical survey.[9] It is true enough that the subject is now a *flourishing* science. But that does not render it immune to conceptual confusions and entanglements. Newtonian kinematics was a flourishing science, but that did not stop Newton from becoming entangled in conceptual confusions over the intelligibility of action at a distance, or from bafflement (not remedied until Hertz) over the nature of force. So too, Sherrington's towering achievement of explaining the integrative action of synapses in the spinal cord, and consequently eliminating, once and for all, the confused idea of a 'spinal soul', was perfectly compatible with conceptual confusions concerning the 'cerebral soul' or mind and its relation to the brain. Similarly, Penfield's extraordinary achievements in identifying functional localization in the cortex, as well as in developing brilliant neurosurgical techniques, were perfectly compatible with extensive confusions about the relation between mind and the brain and about the 'highest brain function' (an idea borrowed from Hughlings Jackson).

In short, conceptual entanglement *can* coexist with flourishing science. This may appear puzzling. If the science can flourish despite such conceptual confusions, why should scientists care about them?—Hidden reefs do not imply that the seas are not navigable, only that they are dangerous. The moot question is how running on these reefs is manifest. Conceptual confusions may be exhibited in different ways and at different points in the investigation. In some cases, the conceptual unclarity may affect neither the cogency of the questions nor the fruitfulness of the experiments, but only the understanding of the results of the experiments and their theoretical implications. So, for example, Newton embarked on the *Optics* in quest of insight into the character of colour. The research was a permanent contribution to science. But his conclusion that 'colours are sensations in the sensorium' demonstrates failure to achieve the kind of understanding he craved. For whatever colours are, they are not 'sensations in the sensorium'. So insofar as Newton cared about *understanding* the results of his research, then he had good reason for caring about the conceptual confusions under which he laboured—for they stood in the way of an adequate understanding.

In other cases, however, the conceptual confusion does not so happily bracket the empirical research. Misguided questions may well render research futile.[10] Rather differently, misconstrual of concepts and conceptual structures will sometimes produce research that is by no means futile, but that fails to show what it was designed to show.[11] In such cases, the science may not be flourishing quite as much as it appears to be. It requires conceptual investigation to locate the problems and to eliminate them.

Are these conceptual confusions *unavoidable*? Not at all. The whole point of writing this book has been to show how to avoid them. But of course, they cannot be avoided while leaving everything else intact. They *can be* avoided—but if they are, then certain kinds of questions will no longer be asked, since they will be recognised as resting on a misunderstanding. As Hertz put it in the wonderful introduction to his *Principles of Mechanics*: 'When these painful contradictions are removed, . . . our minds, no longer vexed, will cease to ask illegitimate questions'. Equally, certain kinds of inferences will no longer be drawn from a given body of empirical research, since it will be realised to have little or no bearing on the matter which it was meant to illuminate, even though it may bear on *something else*.

If there are problematic concepts, can't they be replaced by others that serve the same explanatory function? A scientist is always free to introduce new concepts if he finds existing ones inadequate or insufficiently refined. But our concern in this book is not with the use of new technical concepts. We are concerned with the misuse of old non-technical concepts—concepts of mind and body, of thought and imagination, of sensation and perception, of knowledge and memory, of voluntary movement and of consciousness and self-consciousness. There is nothing inadequate about these concepts relative to the purposes they serve. There is no reason for thinking that they need to be replaced in the contexts that are of concern to us. What is problematic are neuroscientists' misconstruals of them and the misunderstandings consequently engendered. These are remediable by a correct account of the logico-grammatical character of the concepts in question. And this is what we have tried to supply.

Granted that neuroscientists may not be using these common or garden concepts the way the man in the street does, with what right can philosophy claim to correct them? How can philosophy so confidently judge the clarity and coherence of concepts as deployed by competent scientists? How can philosophy be in a position to claim that certain assertions made by sophisticated neuroscientists *make no sense?* We shall resolve such methodological qualms in the following pages. But some initial clarification here may remove some doubts. What truth and falsity is to science, sense and nonsense is to philosophy. Observational and theoretical error result in falsehood; conceptual error results in lack of sense. How can one investigate the bounds of sense? Only by examining the use of words. Nonsense is generated when an expression is used contrary to the rules for its use. The expression in question may be an ordinary, non-technical expression, in which case the rules for its use can be elicited from its standard employment and received explanations of its meaning. Or it may be a technical term of art, in which case the rules for its use must be elicited from the theorist's introduction of the term and the explanations he offers of its stipulated use. Both kinds of term can be misused, and when they are, nonsense ensues—a form of words that is excluded from the language. For either nothing has been stipulated as to what the term means in the aberrant context in question, or this form of words is actually excluded by a rule specifying that there is no such thing as ... (e.g. that there is no such thing as 'east of the North Pole', that this is a form of words that has no use). Nonsense is also commonly generated when an existing expression is given a new, perhaps technical or quasi-technical, use, and the new use is inadvertently crossed with the old, e.g. inferences are drawn from propositions containing the new term which could only licitly be drawn from the use of the old one. It is the task of the conceptual critic to identify such transgressions of the bounds of sense. It is, of course, not enough to show that a certain scientist has used a term contrary to its ordinary use—for he may well be using the term in a new sense. The critic must show that the scientist intends using the term in its customary sense and has not done so, or that he intends using it in a new sense but has inadvertently crossed the new sense with the old. The wayward scientist should, whenever possible, be condemned out of his own mouth.[12]

The final misconception against which we wish to warn is the idea that our reflections are unremittingly negative. All we are concerned with, it might be thought, is criticizing. Our work may appear to a superficial glance to be no more than a destructive undertaking that promises neither assistance nor a new way forward. Worse, it may even appear to be engineering a confrontation between philosophy and cognitive neuroscience. *Nothing could be further from the truth.*

We have written this book in admiration for the achievements of twentieth-century neuroscience, and out of a desire to assist the subject. But the *only* ways a conceptual investigation can assist an empirical subject are by identifying conceptual error (if it obtains) and providing a map that will help prevent empirical researchers from wandering off the highroads of sense. Each of our investigations has two aspects to it. On the one hand, we have tried to identify conceptual problems and entanglements in important current theories of perception, memory, imagination, emotion, and volition. Moreover, we argue that much contemporary writing on the nature of consciousness and self-consciousness is bedevilled by conceptual difficulties. This aspect of our investigations is indeed negative and critical. On the other hand, we have endeavoured, from case to case, to provide a perspicuous representation of the conceptual field of each of the problematic concepts. This is a constructive endeavour. We hope that these conceptual overviews will assist neuroscientists in their reflections antecedent to the design of experiments. However, it cannot be the task of a conceptual investigation to propose empirical hypotheses that might solve the empirical problems faced by scientists. To complain that a philosophical investigation into cognitive neuroscience has not contributed a new neuroscientific theory is like complaining to a mathematician that a new theorem he has proved is not a new physical theory.

PHILOSOPHICAL FOUNDATIONS OF NEUROSCIENCE

An Excerpt from Chapter 3

M. R. BENNETT AND P. M. S. HACKER

3.1 MEREOLOGICAL CONFUSIONS IN COGNITIVE NEUROSCIENCE

Ascribing psychological attributes to the brain

Leading figures of the first two generations of modern brain-neuroscientists were fundamentally Cartesian. Like Descartes, they distinguished the mind from the brain and ascribed psychological attributes to the mind. The ascription of such predicates to human beings was, accordingly, derivative—as in Cartesian metaphysics. The third generation of neuroscientists, however, repudiated the dualism of their teachers. In the course of explaining the possession of psychological attributes by human beings, they ascribed such attributes not to the mind but to the brain or parts of the brain.

Neuroscientists assume that the brain has a wide range of cognitive, cogitative, perceptual and volitional capacities. Francis Crick asserts that

What you see is not what is *really* there; it is what your brain *believes* is there. . . . Your brain makes the best interpretation it can according to its previous experience and the limited and ambiguous information provided by your eyes. . . . The brain combines the information provided by the many distinct features of the visual scene (aspects of shape, colour, movement, etc.) and settles on the most plausible interpretation of all these various clues taken together. . . . What the brain has to build up is a many-levelled interpretation of the visual scene . . . [Filling-in] allows the brain to guess a complete picture from only partial information—a very useful ability.[1]

So the brain *has experiences, believes* things, *interprets* clues *on the basis of information* made available to it and *makes guesses*. Gerald Edelman holds that structures within the brain 'categorize, discriminate, and recombine the various brain activities occurring in different kinds of global mappings', and that the brain 'recursively relates semantic to phonological sequences and then generates syntactic correspondences, not from preexisting rules, but by treating rules developing in memory as objects for conceptual manipulation'.[2] Accordingly the brain *categorizes*, indeed, it 'categorizes its own activities (particularly its perceptual categorizations)' and *conceptually manipulates rules*. Colin Blakemore argues that

We seem driven to say that such neurons [as respond in a highly specific manner to, e.g., line orientation] have knowledge. They have intelligence, for they are able to estimate the probability of outside events—events that are important to the animal in question. And the brain gains its knowledge by a process analogous to the inductive reasoning of the classical scientific method. Neurons present arguments to the brain based on the specific features that they detect, arguments on which the brain constructs its hypothesis of perception.[3]

So the brain *knows* things, *reasons* inductively, *constructs hypotheses* on the basis of arguments, and its constituent neurons are *intelligent*, can *estimate probabilities*, and *present arguments*. J.Z. Young shared much the same view. He argued that 'we can regard all seeing as a continual search for the answers to questions posed by the brain. The signals from the retina constitute "messages" conveying these answers. The brain then uses this information to construct a suitable hypothesis about what is there.'[4] Accordingly, the brain *poses questions, searches for answers*, and *constructs hypotheses*. Antonio Damasio claims that 'our brains can often decide well, in seconds, or minutes, depending on the time frame we set as appropriate for the goal we want to achieve, and if they can do so, they must do the marvellous job with more than just pure reason'[5], and Benjamin Libet suggests that 'the brain "decides" to initiate or, at least, to prepare to initiate the act before there is any reportable subjective awareness that such a decision has taken place.'[6] So brains *decide*, or at least "decide", and *initiate voluntary action*.

Psychologists concur. J.P. Frisby contends that 'there must be a symbolic description in the brain of the outside world, a description cast in symbols which stand for the various aspects of the world of which sight makes us aware.'[7] So there are *symbols in the brain*, and the brain *uses*, and presumably *understands, symbols*. Richard Gregory conceives of seeing as 'probably the most sophisticated of all the brain's activities: calling upon its stores of memory data; requiring subtle classifications, comparisons and logical decisions for sensory data to become perception.'[8] So the brain *sees, makes classifications, comparisons*, and *decisions*. And cognitive scientists think likewise. David Marr held that 'our brains must somehow be capable of *representing . . . information*. . . . The study of vision must therefore include . . . also an inquiry into the nature of the *internal representations* by which we *capture this information* and *make it available as a basis for decisions* about our thoughts and actions.'[9] And Philip Johnson-Laird suggests that the brain 'has access to a partial model of its own capabilities' and has the 'recursive machinery to embed models within models'; consciousness, he contends, 'is the property of a class of parallel algorithms'.[10]

Questioning the intelligibility of ascribing psychological attributes to the brain

With such broad consensus on the correct way to think about the functions of the brain and about explaining the causal preconditions for human beings to possess and exercise their natural powers of thought and perception, one is prone to be swept along by enthusiastic announcements—of new fields of knowledge conquered, new mysteries unveiled.[11] But we should take things slowly, and pause for thought. We know what it is for human beings to experience things, to see things, to know or believe things, to make decisions, to interpret equivocal data, to guess and form hypotheses. We understand what it is for people to reason inductively, to estimate probabilities, to present arguments, to classify and categorize the things they encounter in their experience. We pose questions and search for answers, using a symbolism, namely our language, in terms of which we represent things. But do we know what it is for *a brain* to see or hear, for *a brain* to have experiences, to know or believe something? Do we have any conception of what it would be for *a brain* to make a decision? Do we grasp what it is for a brain (let alone a neuron) *to reason* (no matter whether inductively or deductively), to *estimate probabilities*, to *present arguments*, to *interpret data* and to *form hypotheses* on the basis of its interpretations? We can observe whether a person sees something or other—we look at his behaviour and ask him questions. But what would it be to observe whether a brain sees something—as opposed to observing the brain of *a person* who sees something. We recognize when a person asks a question and when another answers it. But do we have any conception of what it would be for a brain to ask a question or answer one? These are all attributes of human beings. Is it a new *discovery* that brains also engage in such human activities? Or is it a linguistic innovation, introduced by neuroscientists, psychologists and cognitive scientists, extending the ordinary use of these psychological expressions for good theoretical reasons? Or, more ominously, is it a conceptual confusion? Might it be the case that there is simply *no such thing* as the brain's thinking or knowing, seeing or hearing, believing or guessing, possessing and using information, constructing hypotheses, etc., i.e. that these forms of words make no sense? But if there is no such thing, why have so many distinguished scientists thought that these phrases, thus employed, do make sense?

Whether psychological attributes can intelligibly be ascribed to the brain is a philosophical, and therefore a conceptual, question, not a scientific one

The question we are confronting is a philosophical question, not a scientific one. It calls for conceptual clarification, not for experimental investigation. One cannot investigate experimentally whether brains do or do not think, believe, guess, reason, form hypotheses, etc. until one knows what it would be for a brain to do so, i.e. until we are clear about the meanings of these phrases and know what (if anything) *counts* as a brain's doing so and what sort of evidence supports the ascription of such attributes to the brain. (One cannot look for the poles of the earth until one knows what a pole is, i.e. what the expression 'pole' means, and also what counts as finding a pole of the earth. Otherwise, like Winnie-the-Pooh, one might embark on an expedition to the East Pole.) The moot question is: does it make sense to ascribe such attributes to the brain? Is there any such thing as a brain's thinking, believing, etc. (Is there any such thing as the East Pole?)

In the *Philosophical Investigations*, Wittgenstein made a profound remark that bears directly on our concerns. '*Only of a human being and what resembles (behaves like) a living human being can one say: it has sensations; it sees, is blind; hears, is deaf; is conscious or unconscious.*'[12] This epitomizes the conclusions we shall reach in our investigation. Stated with his customary terseness, it needs elaboration, and its ramifications need to be elucidated.

The point is not a factual one. It is not a matter of fact that only human beings and what behaves like human beings can be said to be the subject of these psychological predicates. If it were, then it might indeed be a discovery, recently made by neuroscientists, that brains too see and hear, think and believe, ask and answer questions, form hypotheses and make guesses on the basis of information. Such a discovery would, to be sure, show that it is not only of a human being and what behaves like a human being that one can say such things. This would be astonishing, and we should want to hear more. We should want to know what the evidence for this remarkable discovery was. But, of course, it is not like this. The ascription of psychological attributes to the brain is not warranted by a neuroscientific discovery that shows that contrary to our previous convictions, brains do think and reason, just as we do ourselves. The neuroscientists, psychologists and cognitive scientists who adopt these forms of description have not done so as a result of *observations* which show that brains think and reason. Susan Savage-Rambaugh has produced striking evidence to show that bonobo chimpanzees, appropriately trained and taught, can ask and answer questions, can reason in a rudimentary fashion, give and obey orders, and so on. The evidence lies in their behaviour—in what they do (including how they employ symbols) in their interactions with us. This was indeed very surprising. For no one thought that such capacities could be acquired by apes. But it would be absurd to think that the ascription of cognitive and cogitative attributes to the brain rests on comparable evidence. It would be absurd because we do not even know what would show that the brain has such attributes.

The misascription of psychological attributes to the brain is a degenerate form of Cartesianism Why then was this form of description, and the attendant forms of explanation that are dependent upon it, adopted *without argument or reflection?* We suspect that the answer is—as a result of an unthinking adherence to a mutant form of Cartesianism. It was a characteristic feature of Cartesian dualism to ascribe psychological predicates to the mind, and only derivatively to the human being. Sherrington and his pupils Eccles and Penfield cleaved to a form of dualism in their reflections on the relationship between their neurological discoveries and human perceptual and cognitive capacities. Their successors rejected the dualism—quite rightly. But the predicates which dualists ascribe to the immaterial mind, the third generation of brain neuroscientists applied unreflectively to the brain instead. It was no more than an apparently innocuous corollary of rejecting the two-substance dualism of Cartesianism in neuroscience. These scientists proceeded to explain human perceptual and cognitive capacities and their exercise by reference to the brain's exercise of *its* cognitive and perceptual capacities.

The ascription of psychological attributes to the brain is senseless

It is our contention that this application of psychological predicates to the brain *makes no sense*. It is not that as a matter of fact brains do not think, hypothesize and decide, see and hear, ask and answer questions, rather, it makes no sense to ascribe such predicates *or their negations* to the brain. The brain neither sees *nor is it blind*—just as sticks and stones are not awake, *but they are not asleep either*. The brain does not hear, but it is not deaf, any more than trees are deaf. The brain makes no decisions, but neither is it is indecisive. Only what *can* decide, can be indecisive. So too, the brain cannot be conscious, only the living creature whose brain it is can be conscious—or unconscious. *The brain is not a logically appropriate subject for psychological predicates.* Only a human being and what *behaves* like one can intelligibly and literally be said to see or be blind, hear or be deaf, ask questions or refrain from asking.

Our point, then, is a conceptual one. It makes no sense to ascribe psychological predicates (or their negations) to the brain, save metaphorically or metonymically. The resultant combination of words does not say something that is false, rather it says nothing at all, for it lacks sense. Psychological predicates are predicates that apply essentially to the whole living animal, not to its parts. It is not the eye (let alone the brain) that sees, but *we* see *with* our eyes (and we do not see *with* our brains, although without a brain functioning normally in respect of the visual system, we would not see). So too, it is not the ear that hears, but the animal whose ear it is. The organs of an animal are parts of the animal, and psychological predicates are ascribable to the whole animal, not to its constituent parts.

Neuroscientists' ascription of psychological attributes to the brain may be termed 'the mereological fallacy' in neuroscience

Mereology is the logic of part/whole relations. The neuroscientists' mistake of ascribing to the constituent *parts* of an animal attributes that logically apply only to the *whole* animal we shall call '*the mereological fallacy*' in neuroscience.[13] The principle that psychological predicates which apply only to human beings (or other animals) as wholes cannot intelligibly be applied to their parts, such as the brain, we shall call '*the mereological principle*' in neuroscience.[14] Human beings, but not their brains, can be said to be thoughtful or to be thoughtless; animals, but not their brains, let alone the hemispheres of their brains, can be said to see, hear, smell and taste things; people, but not their brains, can be said to make decisions or to be indecisive.

It should be noted that there are many predicates that *can* apply both to a given whole (in particular a human being) and to its parts, and whose application to the one may be inferred from its application to the other. A man may be sunburnt and his face may be sunburnt; he may be cold all over, so his hands will be cold too. Similarly, we sometimes extend the application of a predicate from a human being to parts of the human body, e.g. we say that a man gripped the handle, and also that his hand gripped the handle, that he slipped and that his foot slipped. Here there is nothing logically awry. But psychological predicates apply paradigmatically to the *human being (or animal) as a whole*, and *not* to the body and its parts. There are a few exceptions, e.g. the application of the verbs of sensation such as 'to hurt', to parts of the body, e.g. 'My hand hurts', 'You are hurting my hand'.[15] But the range of psychological predicates that are our concern, i.e. those that have been invoked by neuroscientists, psychologists and cognitive scientists in their endeavours to explain human capacities and their exercise, have no literal application to parts of the body. In particular they have no intelligible application to the brain.

3.2 METHODOLOGICAL QUALMS

Methodological objections to the accusation that neuroscientists are guilty of a mereological fallacy

If a person ascribes a predicate to an entity to which the predicate in question logically could not apply, and this is pointed out to him, then it is only to be expected that he will indignantly insist that he didn't 'mean it like that'. After all, he may say, since a nonsense is a form of words that says nothing, that fails to describe a possible state of affairs, he obviously did not *mean* a nonsense— one *cannot* mean a nonsense, since there is nothing, as it were, to mean. So his words must not be taken to have their ordinary meaning. The problematic expressions were perhaps used in a special sense, and are really merely homonyms; or they were analogical *extensions* of the customary use—as is indeed common in science; or they were used in a metaphorical or figurative sense. If these escape routes are available, then the accusation that neuroscientists fall victim to the mereological fallacy is unwarranted. Although they make use of the same psychological vocabulary as the man in the street, they are using it in a different way. So objections to neuroscientists' usage based upon the ordinary use of these expressions are irrelevant.

Things are not that straightforward, however. Of course, the person who misascribes a predicate in the manner in question does not intend to utter a form of words that lacks sense. But that he did not mean *to utter* a nonsense does not ensure that he did not do so. Although he will naturally insist that he 'didn't mean it like that', that the predicate in question was not being used in its customary sense, his insistence is not the final authority. The final authority in the matter is *his own reasoning*. We must look at the consequences he draws from his own words—and it is his inferences that will show whether he was using the predicate in a new sense or misusing it. If he is to be condemned, it must be out of his own mouth.

So, let us glance at the proposed escape routes that are intended to demonstrate that neuroscientists and cognitive scientists are not guilty of the errors of which we have accused them.

First objection (Ullman): the psychological predicates thus used are homonyms of ordinary psychological predicates, and have a different, technical, meaning First, it might be suggested that neuroscientists are in effect employing homonyms, which mean something altogether different. There is nothing unusual, let alone amiss, in scientists introducing a new way of talking under the pressure of a new theory. If this is confusing to benighted readers, the confusion can easily be resolved. Of course, brains do not literally think, believe, infer, interpret or hypothesize, they think*, believe*, infer*, interpret* or hypothesize*. They do not have or construct symbolic representations, but symbolic representations*.[16]

Second objection (Gregory): the psychological predicates thus used are analogical extensions of the ordinary expressions Secondly, it might be suggested that neuroscientists are extending the ordinary use of the relevant vocabulary by analogy—as has often been done in the history of science, for example in the analogical extension of hydrodynamics in the theory of electricity. So to object to the ascription of psychological predicates to the brain on the grounds that in ordinary parlance such predicates are applicable only to the animal as a whole would be to display a form of *semantic inertia.*[17]

Third objection (Blakemore): neuroscientists' ascription of psychological attributes to the brain is figurative or metaphorical, since they know perfectly well that the brain does not think or use maps

Finally, it might be argued that neuroscientists do not *really* think that the brain reasons, argues, asks and answers questions just as we do. They do not really believe that the brain interprets clues, makes guesses, or contains symbols which describe the outside world. And although they talk of there being 'maps' in the brain and of the brain's containing 'internal representations', they are not using these words in their common or vulgar sense. This is figurative and metaphorical speech—sometimes even poetic licence.[18] Neuroscientists, therefore, are not in the least misled by such ways of speaking—they know perfectly well what they mean, but lack the words to say it save metaphorically or figuratively.

Reply to the objection that neuroscientists are using the psychological vocabulary in a special technical sense

With regard to the misuse of the psychological vocabulary involved in ascribing psychological predicates to the brain, all the evidence points to the fact that neuroscientists are not using these terms in a special sense. Far from being new homonyms, the psychological expressions they use are being invoked in their customary sense, otherwise the neuroscientists would not draw the inferences from them which they do draw. When Crick asserts that 'what you see is not what is *really* there; it is what your brain *believes* is there ... ' it is important that he takes 'believes' to have its normal connotations—that it does not mean the same as some novel term 'believes*'. For it is part of Crick's tale that the belief is the outcome of *an interpretation* based on previous *experience* and *information* (and not the outcome of an interpretation* based on previous experience* and information*). When Semir Zeki remarks that the acquisition of knowledge is a 'primordial function of the brain'[19], he means knowledge (not knowledge*)—otherwise he would not think that it is the task of future neuroscience to solve the problems of epistemology (but only, presumably, of epistemology*). Similarly, when Young talks of the brain's containing knowledge and information, which is encoded in the brain 'just as knowledge can be recorded in books or computers'[20], he means knowledge (not knowledge*)—since it is knowledge and information (not knowledge* and information*) that can be recorded in books and computers. When Milner, Squire and Kandel talk of 'declarative memory', they explain that this phrase signifies 'what is ordinarily meant by the term "memory"'[21], but then go on to declare that such memories (not memories*) are 'stored in the brain'. That presupposes that it makes sense to speak of storing memories (in the ordinary sense of the word) *in the brain.*[22]

Reply to Ullman: David Marr on 'representations'

The accusation of committing the mereological fallacy cannot be that easily rebutted. But Simon Ullman may appear to be on stronger grounds when it comes to talk of internal representations and symbolic representations (as well as maps) *in the brain.* If 'representation' does not mean what it ordinarily does, if 'symbolic' has nothing to do with symbols, then it may indeed be innocuous to speak of there being internal, symbolic representations in the brain. (And if 'maps' have nothing to do with atlases, but only with *mappings*, then it may also be innocuous to speak of there being maps in the brain.) It is extraordinarily ill-advised to multiply homonyms, but it need involve no conceptual incoherence, *as long as the scientists who use these terms thus do not forget that the terms do not have their customary meaning.* Unfortunately, they typically do forget this and *proceed to cross the new use with the old*, generating incoherence. Ullman, defending Marr, insists (perfectly correctly) that certain brain events can be viewed as representations* of depth or orientation or reflectance[23], i.e. that one can correlate certain neural firings with features in the visual field (denominating the former 'representations*' of the latter). But it is evident that this is not all that Marr meant. He claimed that numeral systems (Roman or Arabic numerals, binary notation) are representations. However, such notations have nothing to do with causal correlations, but with representational conventions. He claimed that 'a representation for shape would be a formal scheme for describing some aspects of shape, together with rules that specify how the scheme is applied to any particular shape'[24], that a formal scheme is 'a set of symbols with rules for putting them together'[25], and that 'a representation, therefore, is not a foreign idea at all—we all use representations all the time. However, the notion that one can capture some aspect of reality by making a description of it using a symbol and that to do so can be useful seems to me to be a powerful and fascinating idea'.[26] But the sense in which we 'use representations all the time', *in which representations are rule-governed symbols, and in which they are used for describing things,* is the *semantic* sense of 'representation'—not a new homonymical causal sense. Marr has fallen into a trap of his own making.[27] He in effect conflates Ullman's representations*, that *are* causal correlates, with representations, that are symbols or symbol systems with a syntax and meaning determined by conventions.

Reply to Ullman: Young on 'maps' and Frisby on 'symbolic representations'

Similarly, it would be misleading, but otherwise innocuous, to speak of maps in the brain when what is meant is that certain features of the visual field can be mapped onto the firings of groups of cells in the 'visual' striate cortex. But then one cannot go on to say, as Young does, that the brain makes use of its maps in formulating its hypotheses about what is visible. So too, it would be innocuous to speak of there being symbolic representations in the brain, as long as 'symbolic' has nothing to do with semantic meaning, but signifies only 'natural meaning' (as in 'smoke means fire'). But then one cannot go on to say, as Frisby does, that 'there must be a symbolic description in the brain of the outside world, a description cast in symbols which stand for the various aspects of the world of which sight makes us aware'.[28] For this use of 'symbol' is evidently semantic. For while smoke means fire, in as much as it is a sign *of* fire (an inductively correlated indication), it is not a sign *for* fire. Smoke rising from a distant hillside is not a description of fire cast in symbols, and the firing of neurons in the 'visual' striate cortex is not a symbolic description of objects in the visual field, even though a neuroscientist may be able to infer facts about what is visible to an animal from his knowledge of what cells are firing in its 'visual' striate cortex. The firing of cells in V1 may be signs of a figure with certain line orientations in the animal's visual field, but they do not *stand for* anything, they are not symbols, and they do not describe anything.

Reply to the second objection that in ascribing psychological attributes to the brain , neuroscientists are not committing the mereological fallacy, but merely extending the psychological vocabulary analogically

The thought that neuroscientific usage, far from being conceptually incoherent, is innovative, extending the psychological vocabulary in novel ways, might seem to offer another way of sidestepping the accusation that neuroscientists' descriptions of their discoveries commonly transgress the bounds of sense. It is indeed true that analogies are a source of scientific insight. The hydrodynamical analogy proved fruitful in the development of the theory of electricity, even though electrical current does not flow in the same sense as water flows and an electrical wire is not a kind of pipe. The moot question is whether the application of the psychological vocabulary to the brain is to be understood as analogical.

The prospects do not look good. The application of psychological expressions to the brain is not part of a complex theory replete with functional, mathematical relationships expressible by means of quantifiable laws as are to be found in the theory of electricity. Something much looser seems to be needed. So, it is true that psychologists, following Freud and others, have extended the concepts of belief, desire and motive in order to speak of *unconscious* beliefs, desires and motives. When these concepts undergo such analogical extension, something new stands in need of explanation. The newly extended expressions no longer admit of the same combinatorial possibilities as before. They have a different, importantly related, meaning, and one which requires explanation. The relationship between a (conscious) belief and an unconscious belief, for example, is not akin to the relationship between a visible chair and an occluded chair—it is not 'just like a conscious belief only unconscious', but more like the relationship between $\sqrt{1}$ and $\sqrt{-1}$. But when neuroscientists such as Sperry and Gazzaniga speak of the left hemisphere making choices, of its generating interpretations, of its knowing, observing and explaining things—it is clear from the sequel that these psychological expressions have not been given a new meaning. Otherwise it would not be said that a hemisphere of the brain is 'a conscious system in its own right, perceiving, thinking, remembering, reasoning, willing and emoting, *all at a characteristically human level*'.[29]

It is not semantic inertia that motivates our claim that neuroscientists are involved in various forms of conceptual incoherence. It is rather the acknowledgement of the requirements of the logic of psychological expressions. Psychological predicates are predicable only of a whole animal, not of its parts. No conventions have been laid down to determine what is to be meant by the ascription of such predicates to a part of an animal, in particular to its brain. So the application of such predicates to the brain or the hemispheres of the brain transgresses the bounds of sense. The resultant assertions are not false, for to say that something is false, we must have some idea of what it would be for it to be true—in this case, we should have to know what it would be for the brain to think, reason, see and hear, etc. and to have found out that as a matter of fact the brain does not do so. But we have no such idea, and these assertions are not false. Rather, the sentences in question lack sense. This does not mean that they are silly or stupid. It means that no sense has been assigned to such forms of words, and that accordingly they say nothing at all, even though it looks as if they do.

Reply to the third objection (Blakemore)
that applying psychological predicates
to the brain is merely metaphorical

The third methodological objection was raised by Colin Blakemore. Of Wittgenstein's remark that 'only of a living human being and what resembles (behaves like) a living human being can one say: it has sensations; it sees; is blind; hears; is deaf; is conscious or unconscious', Blakemore observes that it 'seems trivial, maybe just plain wrong'. Addressing the accusation that neuroscientists' talk of there being 'maps' in the brain is pregnant with possibilities of confusion (since all that can be meant is that one can map, for example, aspects of items in the visual field onto the firing of cells in the 'visual' striate cortex), Blakemore notes that there is overwhelming evidence for 'topographic patterns of activity' in the brain.

> Since Hughlings Jackson's time, the concept of functional sub-division and topographic representation has become a *sine qua non* of brain research. The task of charting the brain is far from complete but the successes of the past make one confident that each part of the brain (and especially the cerebral cortex) *is* likely to be organized in a spatially ordered fashion. Just as in the decoding of a cipher, the translation of Linear B or the reading of hieroglyphics, all that we need to recognize the order in the brain is a set of rules—rules that relate the activity of the nerves to events in the outside world or in the animal's body.[30]

To be sure, the term 'representation' here merely signifies systematic causal connectedness. That is innocuous enough. But it must not be confused with the sense in which a sentence of a language can be said to represent the state of affairs it describes, a map to represent that of which it is a map, or a painting to represent that of which it is a painting. Nevertheless, such ambiguity in the use of 'representation' is perilous, since it is likely to lead to a confusion of the distinct senses. Just how confusing it can be is evident in Blakemore's further observations:

> Faced with such overwhelming evidence for topographic patterns of activity in the brain it is hardly surprising that neurophysiologists and neuroanatomists have come to speak of the brain having *maps*, which are thought to play an essential part in the representation and interpretation of the world by the brain, just as the maps of an atlas do for the reader of them. The biologist J.Z. Young writes of the brain having a language of a pictographic kind: 'What goes on in the brain must provide a faithful representation of events outside it, and the arrangements of the cells in it provide a detailed model of the world. It communicates meanings by topographical analogies'.[31] But is there a danger in the metaphorical use of such terms as 'language', 'grammar', and 'map' to describe the properties of the brain? . . . I cannot believe that any neurophysiologist believes that there is a ghostly cartographer browsing through the cerebral atlas. Nor do I think that the employment of common language words (such as map, representation, code, information and even language) is a conceptual blunder of the kind [imagined]. Such metaphorical imagery is a mixture of empirical description, poetic licence and inadequate vocabulary.[32]

Whether there is any danger in a metaphorical use of words depends on how clear it is that it is merely metaphorical, and on whether the author remembers that that *is* all it is. Whether neuroscientists' ascriptions to the brain of attributes that can be applied literally only to an animal as a whole is actually merely metaphorical (metonymical or synecdochical) is very doubtful. *Of course*, neurophysiologists do not think that there is a 'ghostly cartographer' browsing through a cerebral atlas—but they do think that the brain *makes use* of the maps. According to Young, the brain *constructs hypotheses*, and it does so *on the basis* of this 'topographically organized representation'.[33] The moot question is: what inferences do neuroscientists draw from their claim that there are maps or representations in the brain, or from their claim that the brain contains information, or from talk (J.Z. Young's talk) of 'languages of the brain'? These alleged metaphorical uses are so many banana-skins in the pathway of their user. He *need not* step on them and slip, but he probably will.

Blakemore's confusion   Just how easy it is for confusion to ensue from what is alleged to be harmless metaphor is evident in the paragraph of Blakemore quoted above. For while it may be harmless to talk of 'maps', i.e. of *mappings* of features of the perceptual field onto topographically related groups of cells that are systematically responsive to such features, it is anything but harmless to talk of such 'maps' as playing 'an essential part in the *representation* and *interpretation* of the world by the brain, just as the maps of an atlas do for the reader of them' (our italics). In the first place, it is not clear what sense is to be given to the term 'interpretation' in this context. For it is by no means evident what could be meant by the claim that the topographical relations between groups of cells that are systematically related to features of the perceptual field play an essential role in the brain's *interpreting* something. To interpret, literally speaking, is to explain the meaning of something, or to take something that is ambiguous to have one meaning rather than another. But it makes no sense to suppose that the brain *explains* anything, or that it apprehends something as *meaning* one thing rather than another. If we look to J.Z. Young to find out what he had in mind, what we find is the claim that it is on the basis of such maps that the brain 'constructs hypotheses and programs'—and this only gets us deeper into the morass.

More importantly, whatever sense we can give to Blakemore's claim that 'brain-maps' (which are not actually maps) play an essential part in the brain's 'representation and interpretation of the world', it *cannot* be *'just as the maps of an atlas do for the reader of them'*. For a map is a pictorial representation, made in accordance with conventions of mapping and rules of projection. Someone who can read an atlas must know and understand these conventions, and read off, from the maps, the features of what is represented. But the 'maps' in the brain are not maps, in this sense, at all. The brain is not akin to the reader of a map, since it cannot be said to know any conventions of representations or methods of projection or to read anything off the topographical arrangement of firing cells in accordance with a set of conventions. For the cells are not arranged in accordance with conventions at all, and the correlation between their firing and features of the perceptual field is not a conventional but a *causal* one.[34]

PHILOSOPHICAL FOUNDATIONS OF NEUROSCIENCE

An Excerpt from Chapter 10

M. R. BENNETT AND P. M. S. HACKER

10.3 QUALIA

Qualia conceived of as the qualitative character of experience—the philosophers' conception

The temptation to extend the concept of consciousness to encompass the whole domain of 'experience' was greatly strengthened by philosophers' misconceived introduction of the notion of qualia. Neuroscientists unfortunately picked up this aberrant idea and the misconceptions associated with it. The term 'qualia' was introduced to signify the alleged 'qualitative character of experience'. Every experience, it is claimed, has a distinctive qualitative character. Qualia, Ned Block holds, 'include the ways it feels to see, hear and smell, the way it feels to have a pain; more generally, what it's like to have mental states. Qualia are experiential properties of sensations, feelings, perceptions and ... thoughts and desires as well.'[1] Similarly, Searle argues that 'Every conscious state has a certain qualitative feel to it, and you can see this if you consider examples. The experience of tasting beer is very different from hearing Beethoven's Ninth Symphony, and both of those have a different qualitative character from smelling a rose or seeing a sunset. These examples illustrate the different qualitative features of conscious experiences.'[2] Like Block, Searle too holds that thinking has a special qualitative feel to it: 'There is something it is like to think that two plus two equals four. There is no way to describe it except by saying that it is the character of thinking consciously "two plus two equals four".'[3] The subject matter of an investigation of consciousness, Chalmers suggests, 'is best characterized as "the subjective quality of experience"'. A mental state is conscious, he claims, 'if it has a *qualitative feel*—an associated quality of experience. These qualitative feels are also known as phenomenal qualities, or *qualia* for short. The problem of explaining these phenomenal qualities is just the problem of explaining consciousness.'[4] He too takes the view that thinking is an experience with a qualitative content: 'When I think of a lion, for instance, there seems to be a whiff of leonine quality to my phenomenology: what it is like to think of a lion is subtly different from what it is like to think of the Eiffel tower.'[5]

Neuroscientists follow the philosophers

Neuroscientists have gone along with the notion of qualia. Ian Glynn contends that 'Although qualia are most obviously associated with sensations and perceptions, they are also found in other mental states, such as beliefs, desires, hopes, and fears, during conscious episodes of these states.'[6] Damasio states that 'Qualia are the simple sensory qualities to be found in the blueness of the sky or the tone of a sound produced by a cello, and the fundamental components of the images [of which perception allegedly consists] are thus made up of qualia.'[7] Edelman and Tononi hold that 'each differentiable conscious experience represents a different quale, whether it is primarily a sensation, an image, a thought, or even a mood ... '[8], and go on to claim that 'the problem of qualia' is 'perhaps the most daunting problem of consciousness'.

Explaining the qualitative character of experience in terms of there being something it is like to have it

The subjective or qualitative feel of a conscious experience is in turn characterized in terms of there being *something it is like* for an organism to have the experience. What it is like *is* the subjective character of the experience. 'An experience or other mental entity is "phenomenally conscious"', the *Routledge Encyclopaedia of Philosophy* tells us, 'just in case there is something it is like for one to have it.'[9] 'Conscious states are qualitative', Searle explains, 'in the sense that for any conscious state . . . there is something that it qualitatively feels like to be in that state.'[10] The idea, and the mesmerizing turn of phrase 'there is something which it is like', derive from a paper by the philosopher Thomas Nagel entitled 'What is it like to be a bat?'. Nagel argued that 'the fact that an organism has conscious experience *at all* means, basically, that there is something it is like to *be* that organism. . . . Fundamentally an organism has conscious mental states if and only if there is something it is like to *be* that organism—something it is like *for* the organism.'[11] This, i.e. what it is like for the organism, is the subjective character or quality of experience.

Nagel's explanation of consciousness in terms of there being something it is like . . .

If we take for granted that we understand the phrase 'there is something which it is like' thus used, then it seems that Nagel's idea gives us a handle on the concept of a conscious creature and on the concept of a conscious experience:

(1) *A creature is conscious or has conscious experience if and only if there is something which it is like for the creature to be the creature it is.*

(2) *An experience is a conscious experience if and only if there is something which it is like for the subject of the experience to have it.*

So, there is something which it is like for a bat to be a bat (although, Nagel claims, we cannot imagine *what* it is like), and there is something which it is like for us to be human beings (and, he claims, we all know what it is like for us to be us).

It is important to note that the phrase 'there is something *which it is like* for a subject to have experience E' does *not* indicate *a comparison*. Nagel does not claim that to have a given conscious experience *resembles* something (e.g. some other experience), but rather that there is something which it is like *for the subject* to have it, i.e. 'what it is like' is intended to signify 'how it is for the subject himself'.[12] It is, however, striking that Nagel never tells us, with regard to even one experience, what it is like for anyone to have it. He claims that the qualitative character of the experiences of other species may be beyond our ability to conceive. Indeed, the same may be true of the experiences of other human beings. 'The subjective character of the experience of a person deaf and blind from birth is not accessible to me, for example, nor is mine to him.' But we know what it is like to be us, 'and while we do not possess the vocabulary to describe it adequately, its subjective character is highly specific, and in some respects describable in terms that can be understood only by creatures like us.'[13]

Philosophers and neuroscientists concur

Philosophers and neuroscientists have gone along with this idea. It seems to them to capture the essential nature of conscious beings and conscious experience. Thus Davies and Humphries contend that, 'while there is nothing that it is like to be a brick, or an ink-jet printer, there is, presumably, something it is like to be a bat, or a dolphin, and there is certainly something it is like to be a human being.

A system—whether a creature or artefact—is conscious just in case there is something it is like to be that system.'[14] Edelman and Tononi agree that 'We know what it is like to be us, but we would like to explain why we are conscious at all, why there is "something" it is like to be us—to explain how subjective experiential qualities are generated.'[15] And Glynn holds that with respect to our experiences, e.g. of smelling freshly ground coffee, hearing an oboe playing, or seeing the blue of the sky, 'we know what it is like to have these experiences only by having them or by having had them. . . . Just as it feels like something to smell freshly ground coffee, so it can feel like something (at least intermittently) to believe that . . . , or to desire that . . . , or to fear that . . . '

Qualia, then, are conceived to be the qualitative characteristics of 'mental states' or of 'experiences', the latter pair of categories being construed to include not only perception, sensation and affection, but also desire, thought and belief. For every 'conscious experience' or 'conscious mental state', there is something which it is like for the subject to have it or to be in it. This something is a quale—a 'qualitative feel'. 'The problem of explaining these phenomenal qualities', Chalmers declares, 'is just the problem of explaining consciousness.'[16]

10.31 'HOW IT FEELS' TO HAVE AN EXPERIENCE

The primary rationale for extending the ordinary concept of consciousness

One reason given for extending the concept of consciousness beyond its legitimate conservative boundaries was that what is distinctive, remarkable, indeed mysterious, about experiences is *that there is something which it is like to have them.* An experience, it is argued, *is a conscious* experience just in case there is something which it is like for the subject of the experience to have it. Consciousness, thus conceived, is *defined* in terms of *the qualitative feel of experience.* There is a specific way it feels to see, hear and smell, to have a pain, or indeed 'to have mental states' (Block); every conscious state has a certain *qualitative feel* to it (Searle), and each differentiable conscious experience represents a different quale (Edelman and Tononi). This qualitative feel, unique to every distinguishable experience, is *what it is like for the subject of the experience to have the experience.* Or so it is held.

Our suspicions should be aroused by the odd phrases used to invoke something with which we are all supposed to be utterly familiar. We shall examine 'ways of feeling' first, and there being 'something which it is like' subsequently.

Is there always a way it feels to have a 'conscious experience'?

Is there really *a specific way* it feels to see, hear, smell? One might indeed ask a person who has had his sight, hearing or sense of smell restored 'How does it feel to see (hear, smell) again?' One might expect the person to reply 'It is wonderful', or perhaps 'It feels very strange'. The question concerns the person's attitude towards his exercise of his restored perceptual capacity—so, he finds it wonderful to be able to see again, or strange to hear again after so many years of deafness. In these cases, there is indeed a way it feels to see or hear again, namely wonderful or strange. But if we were to ask a normal person how it feels to see the table, chair, desk, carpet, etc., etc., he would wonder what we were after. There is nothing distinctive about seeing these mundane objects. Of course, seeing the table differs from seeing the chair, desk, carpet, etc., but the difference does not consist in the fact that seeing the desk *feels different* from seeing the chair. Seeing an ordinary table or chair does not evoke *any* emotional or attitudinal reaction whatsoever in normal circumstances. The experiences differ in so far as their objects differ.

One may say, clumsily, that there is a way it feels to have a pain. That is just a convoluted way of saying that there is an answer to the (rather silly) question 'How does it feel to have a pain?', e.g. that it is very unpleasant, or, in some cases, dreadful. So, one

may say that there is a way it feels to have migraine, namely very
unpleasant. That is innocuous, but lends no weight to the general
claim that for every differentiable experience, there is a specific
way it feels to have it. Pains are an exception, since they, by definition, have a negative hedonic tone. Pains are sensations which
are intrinsically disagreeable. Perceiving, however, is not a matter
of having sensations. And perceiving in its various modalities and
with its indefinitely numerous possible objects can often be, but
typically is not, the subject of any affective or attitudinal quality
(e.g. pleasant, enjoyable, horrible) at all, let alone a different one
for each object in each perceptual modality. And for a vast range
of things that can be called 'experiences', there isn't 'a way it
feels' to have them, i.e. there is no answer to the question 'How
does it feel to . . . ?'

One cannot but agree with Searle that the experience of tasting beer is very different from hearing Beethoven's Ninth, and
that both are different from smelling a rose or seeing a sunset, for
perceptual experiences are essentially identified or specified by
their modality, i.e. sight, hearing, taste, smell and tactile perception, and by their objects, i.e. by what they are experiences of.
But to claim that the several experiences have a unique, distinctive *feel* is a different and altogether more questionable claim. It
is more questionable in so far as it is obscure what is *meant*. Of
course, all four experiences Searle cites are, for many people,
normally enjoyable. And it is perfectly correct that the identity
of the pleasure or enjoyment is dependent upon the object of the
pleasure. One cannot derive the pleasure of drinking beer from
listening to Beethoven's Ninth, or the pleasure of seeing a sunset
from smelling a rose. That is a logical, not an empirical, truth, i.e.
it is not that, as a matter of fact, the qualitative 'feel' distinctive
of seeing a sunset differs from the 'feel' distinctive of smelling a
rose—after all, both may be very pleasant. Rather, as a matter of
logic, the pleasure of seeing a sunset differs from the pleasure of
smelling a rose, for the identity of the pleasure depends upon
what it is that pleases. It does not follow that every experience

has a different *qualitative character*, i.e. that there is a specific 'feel' to each and every experience. For, first, most experiences have, in this sense, no qualitative character at all—they are neither agreeable nor disagreeable, neither pleasant nor unpleasant, etc. Walking down the street, we may see dozens of different objects. Seeing a lamp post is a different experience from seeing a post-box—did it have a different 'feel' to it? No; and it didn't have the same 'feel' to it either, for seeing the two objects evoked no response—no 'qualitative feeling' whatsoever was associated with seeing either of them. Second, different experiences which *do* have a qualitative 'feel', i.e. which can, for example, be hedonically characterized, may have the very same 'feel'. What differentiates them is not the way they feel, in as much as the question, 'What did it feel like to V?' (where 'V' specifies some appropriate experience) may have exactly the same answer—for the different experiences may be equally enjoyable or disagreeable, interesting or boring.

The qualitative character of experiences correctly construed

Both having a pain (being in pain) and perceiving whatever one perceives can be called 'experiences'. So can being in a certain emotional state. And so, of course, can engaging in an indefinite variety of activities. Experiences, we may say, are possible subjects of attitudinal predicates, that is, they may be agreeable or disagreeable, interesting or boring, wonderful or dreadful. It is such attributes that might be termed 'the qualitative characters of experiences', not the experiences themselves. So one cannot intelligibly say that seeing red or seeing *Guernica*, hearing a sound or hearing *Tosca*, are 'qualia'. Consequently, when Damasio speaks of the blueness of the sky as being a quale, he is shifting the sense of the term 'quale'—since if the colour of an object is a quale, then qualia are not the qualitative characteristics of experiences at all, but the qualities of objects of experience (or, if one holds colours not to be qualities of objects, then constituents of the contents of perceptual experiences). Similarly, when Edelman and Tononi claim that each differentiable conscious experience represents a different quale, whether it is a sensation, an image, a mood or a thought, they are shifting the sense of the term 'quale'. For it patently does not mean 'the qualitative character of an experience' in the sense we have been investigating. What it does, or is supposed to, mean is something we shall examine shortly (§10.34).

It should be noted that to say that an experience is a subject of an attitudinal predicate is a potentially misleading *façon de parler*. For to say that an experience (e.g. seeing, watching, glimpsing, hearing, tasting this or that, but also walking, talking, dancing, playing games, mountain climbing, fighting battles, painting pictures) had a given qualitative feel to it (e.g. that it was agreeable, delightful, charming, disagreeable, revolting, disgusting) is just to say that the subject of experience, i.e. the person who saw, heard, tasted, walked, talked, danced, etc., found it agreeable, delightful, charming, etc. to do so. So, the qualitative character of an experience E, i.e. how it feels to have that experience, is the subject's affective attitude (what it was like for him) to experiencing E.

To avoid falling into confusion here, we must distinguish four points:

(1) Many experiences are essentially individuated, i.e. picked out, by specifying what they are experiences of.

(2) Every experience is a *possible* subject of positive and negative attitudinal predicates, e.g. predicates of pleasure, interest, attraction. It does not follow, and it is false, that every experience is an *actual* subject of a positive or negative attitudinal predicate.

(3) Distinct experiences, each of which is the subject of an attitudinal attribute, may not be distinguishable by reference to how it feels for the person to have them. Roses have a different smell from lilac. Smelling roses is a different experience from smelling lilac. One cannot get the pleasure of smelling roses from smelling lilac. But the experiences may well be equally agreeable. So, if asked how it feels to smell roses and how it feels to smell lilac, the answer may well be the same, namely 'delightful'.

If that answer specifies the way it felt, then it is obviously false that every distinct experience can be uniquely individuated by its distinctive qualitative character or quale. We must not confuse the qualitative character of the experience with the qualitative character of the object of the experience. It is the latter, not the former, that individuates the experience.

(4) Even if we stretch the concept of experience to include thinking that something is so or thinking of something, what essentially differentiates thinking one thing rather than another is not how it feels or what it feels like to think whatever one thinks. Thinking that 2+2 = 4 differs from thinking that 25 x 25 = 625 and both differ from thinking that the Democrats will win the next election.[17] They differ in as much as they are essentially specified or individuated by their objects. One can think *that* something is thus-and-so or think *of* something or other without any accompanying affective attitude whatsoever—so there need be no 'way it feels' to think thus. A leonine whiff may accompany thinking of lions, of Richard Coeur de Lion, or of Lyons Corner House, but, contrary to Chalmers, to specify the associated whiff is not to characterize *how it feels* to think of such items, let alone uniquely to individuate the thinking. That one associates thinking of one of these with a leonine whiff is no answer to the (curious) question 'How does it feel to think of lions (Richard Coeur de Lion, Lyons Corner House)?', and certainly does not distinguish one's thinking of lions as opposed to thinking of Lyons's or Richard I.

PHILOSOPHICAL FOUNDATIONS
OF NEUROSCIENCE

An Excerpt from Chapter 14:
The Concluding Remarks

M.R. BENNETT AND P.M.S. HACKER

14.5 WHY IT MATTERS

On the question of how it will affect the next experiment
We can imagine a scientist reading our analytical discussions with some bafflement. He might be mildly interested in some of our connective analyses, yet nevertheless puzzled at what seems to be endless logic chopping. 'Does all this really matter?', he might query when he has read our opening discussions. 'After all', he might continue, 'how is this going to affect the next experiment?' We hope that any reader who has followed us thus far will not be tempted to ask this question. For it displays incomprehension.

Whether our analytic reflections do or do not affect the next experiments is *not* our concern. They may or may not—that depends on what experiment is in view, and what the neuroscientist's presuppositions are. It should be obvious, from our foregoing discussions, that, if our arguments are cogent, some experiments might best be abandoned.[1] Others would need to be redesigned.[2] Most may well be unaffected, although the questions addressed might well need to be rephrased, and the results would need to be described in quite different ways than hitherto.[3]

Our concern is with understanding the last experiment

Our concern has not been with the design of the next experiment, but rather with the *understanding* of the last experiment. More generally, conceptual investigations contribute primarily to understanding what is known, and to clarity in the formulation of questions concerning what is not known. It would not matter in the least if our reflections have *no* effect on the next experiment. But they do have considerable effect on the interpretation of the results of previous experiments. And they surely have something to contribute to the asking of questions, to the formulation of questions, and to distinguishing between significant and confused questions. (If we are right, then questions about 'the binding problem', understood as the problem of how the brain forms images, are largely expressions of confusion[4], and much of the debate about mental imagery is misconceived.[5])

Does it matter? If understanding matters, then it matters

Does all this apparent logic chopping, all this detailed discussion of words and their use, *matter*? Does neuroscience really *need* this sort of thing? If the moving spirit behind the neuroscientific endeavour is the desire to understand neural phenomena and their relation to psychological capacities and their exercise, then it matters greatly. For irrespective of the brilliance of the neuroscientist's experiments and the refinement of his techniques, if there is conceptual confusion about his questions or conceptual error in the descriptions of the results of his investigations, then he will not have understood what he set out to understand.

Most contemporary neuroscientists working in the domain of cognitive neuroscience agree that Sir John Eccles's advocacy of a form of dualism was a mistake[6]—and it is a *conceptual confusion* that lies at the heart of Eccles's error. We have tried to demonstrate, by reference to a variety of theories of distinguished contemporary cognitive neuroscientists, that conceptual error, far from being eradicated by a superficial rejection of various forms Cartesian dualism, is widespread. It affects and infects the cogency of the questions addressed, the character of the experiments devised to answer them, the intelligibility of the descriptions of the results of these experiments and the coherence of the conclusions derived from them. And this surely matters both to the understanding of what current neuroscientists *have* achieved, and to the further progress of cognitive neuroscience.

Why it matters to the educated public   It also matters greatly to the educated public. For irrespective of whether certain neuroscientists are confused, there is no question but that the forms of description they employ confuse the lay public. Neuroscientists are understandably eager to communicate the knowledge they have attained over the past decades about the functioning of the brain and to share with the educated public some of the excitement they feel about their subject. That is evident from the flood of books written by numerous distinguished members of the profession. But by speaking about the brain's thinking and reasoning, about one hemisphere's knowing something and not informing the other, about the brain's making decisions without the person's knowing, about rotating mental images in mental space, and so forth, neuroscientists are fostering a form of mystification and cultivating a neuro-mythology that are altogether deplorable. For, first, this does anything but engender the understanding on behalf of the lay public that is aimed at. Secondly, the lay public will look to neuroscience for answers to pseudo-questions that it should not ask and that neuroscience cannot answer. Once the public become disillusioned, they will ignore the important genuine questions neuroscience *can* both ask and answer. And this surely matters.

On the need for conceptual clarity   We have, throughout this book, tried to show that clarity concerning conceptual structures is as important for cognitive neuroscience as clarity about experimental methods. Its great contributions to our understanding of the biological roots of human capacities and their exercise are illuminated, not hindered, by such clarification. For only when the long shadows cast by conceptual confusions are chased away can the achievements of neuroscience be seen aright.

NEUROSCIENCE AND PHILOSOPHY

MAXWELL BENNETT

A Personal Odyssey

When a propagating action potential reaches a synapse at the
end of an axon terminal of a presynaptic neuron, indicated by
the small rectangle in figure 1, it induces the release of neu-
rotransmitter molecules, as shown in the inset of a synapse in the
lower left of figure 1. The transmitter diffuses across a narrow
cleft and binds to receptors in the postsynaptic membrane. Such
binding leads to the opening of channels and often, in turn, to
the generation of action potentials in the postsynaptic neuron.
There are several hundred proteins required for this process (Sieburth et al. 2005). I have spent more than forty years researching
the mechanisms involved in transmission at the synapse (Bennett
2001) and recently began a series of investigations on how networks of synapses operate to fulfill their functions in the brain.
Such networks, consisting of thousands to millions of neurons,

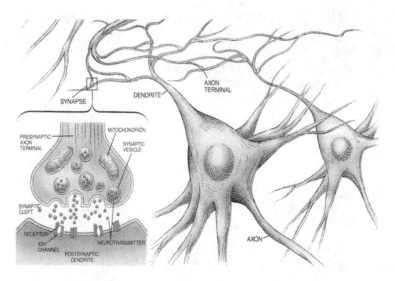

FIGURE I. The axons of two neurons possess processes that terminate in intimate contact with each other at sites called synapses. One of these is boxed, and an enlargement of the synapse within the box is shown in the left corner. A description of the operation of this synapse is given in the text.

each possessing up to about ten thousand synapses, can be found in parts of the brain that must function normally in order to be able to remember a novel event for more than about one minute (the hippocampus), to see (the retina and primary visual cortex V1), and to acquire a variety of motor skills (cerebellum).

The networks of synapses I chose for the initial study were those in the hippocampus (Bennett, Gibson, and Robinson 1994). The general distribution of neuron types and their synapses were first described for the hippocampus by Ramon y Cajal (1904) and are shown on the left in figure 2. The alphabetical letters on the figure refer to different parts of the hippocampus as well as the neuron types (indicated by black ellipsoids and possessing long thin dendritic and axon processes), together with their synaptic connections, as indicated in the legend to the figure. An engineering approach to trying to understand the functioning of the hippocampus involves development of a neural network representation like that shown on the right in figure 2 and described in the legend. Here neuron types are now indicated by circles, their dendrites and axonal processes by straight lines, and synapses by small rectangles. As a young man, having just completed a degree in electrical engineering, I was intrigued by Brindley's (1967) suggestion that some synapses in networks such as these are modifiable. What he meant by this term is that the synapses are able permanently to change their properties following the arrival of an action potential in the axon terminal. Given this possibility, then, it may be that "conditioning and memory mechanisms of the nervous system store information by means of modifiable synapses." Brindley (1969) went on to show how such modifiable synapses in neural network models could "perform many simple learning tasks." Subsequently Marr (1971), also working at Cambridge University, suggested that "the most important characteristic of archicortex (hippocampus) is its ability to perform a simple kind of memorizing task." It was he who first suggested that a recurrent network of collaterals (figure 2) could act as an autoassociative memory if the efficacy of the excitatory synapses were modifiable and if the membrane potentials of the pyramidal neurons were set by inhibitory interneurons that measure the total activity of the network. His suggestion was framed in engineering terms that I and many others found very attractive for further theoretical and experimental research. My colleagues and I followed this general conceptual approach of Brindley and Marr, identifying the conditions under which a neural network representation of the hippocampus like that in figure 2 could work (Bennett, Gibson, and Robinson 1994). We suggested that "the recall of a memory begins with the firing of a set of pyramidal neurons that overlap with the memory to be recalled" and that "the firing of different sets of pyramidal neurons then evolves by discrete synchronous steps" until the stored memory pattern of neurons is retrieved (figure 2).[1] There are, however, two aspects of this kind of engineering approach to understanding the functioning of synaptic networks and therefore of the brain that I found puzzling and these are detailed in the sections that follow.[2]

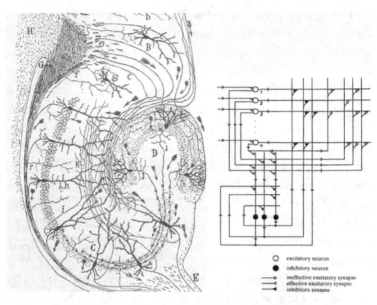

FIGURE 2. On the left is a drawing illustrating neurons and their synaptic connections in the hippocampus after their staining by Ramon y Cajal (figure 479 in Ramon y Cajal 1904). The alphabetical letters on the figure refer to different components of the hippocampus including, for the present purposes: D, dentate gyrus; K, recurrent collaterals from pyramidal cells in the CA3 region of the hippocampus (C) to form synapses on each other as well as projecting to form synapses on pyramidal neurons in the CA1 region of the hippocampus (h).

On the right is a diagram of the basic synaptic network in the CA3 region of the hippocampus, consisting of pyramidal neurons (open circles) and inhibitory interneurons (filled circles). The pyramidal neurons make random connections with each other through their recurrent collaterals. Before learning, these connections are ineffective; after learning, a subset of them becomes effective and in the final evolved state of the network there are excitatory synaptic connections whose strengths are taken to be unity (open triangles) and others whose strengths have remained at zero (open circles). The inhibitory interneurons receive random connections from many pyramidal cells and also from inhibitory neurons. The inhibitory neurons in turn project to pyramidal neurons. The strength of any synapse involving inhibitory interneurons is taken as fixed. The initial state of the system is set by a firing pattern coming onto the pyramidal neurons from either the mossy fiber axons emerging from region D in Ramon y Cajal's drawing or the direct perforant pathway above region D, and this is shown by the lines entering from the left. Once the initial state has been set, the external source is removed. The CA3 recurrent network then updates its internal state cyclically and synchronously.

(continued)

First Major Source of Concern Relating to
Our Understanding of the Function of Cellular
Networks in the Cortex

Given that we know that injury to the hippocampus prevents one from remembering an event for more than about one minute, how are the cells identified in the normal hippocampus that are implicated in our being able to remember, and what are the synaptic relationships between these cells and how do these synapses function? For instance, in any given volume of the hippocampus there are many more glial cells than neurons, and these glial cells come in a variety of types as do neurons. The discovery of propagating and transmitting waves of activity between astrocytic glial cells (Cornell-Bell et al. 1990), although much slower than that between neurons (Bennett, Farnell, and Gibson 2005), introduces considerable complexity into a search for the cellular correlates of memory in the hippocampus. Although such glial waves have been dismissed as unlikely to be relevant in the search for cellular correlates of our psychological attributes (Koch 2004), no experiments have been performed that shows this is the case. Given the physical intimacy between

In the model neural network inhibitory neurons are modeled as rapidly acting linear devices that produce outputs proportional to their inputs; they perform an important regulatory function in the setting of the membrane potentials of the pyramidal neurons. The probability of a neuron firing in a so-called stored memory, which determines the average number of neurons active when a "memory is recalled," can be set at will. "Memories" in this network are allegedly stored at the recurrent collateral synapses using a two-valued Hebbian. Allowance is made in the theory both for the spatial correlations between the learned strengths of the recurrent collateral synapses and temporal correlations between the state of the network and these synaptic strengths. The recall of a memory is conceived of as beginning with the firing of a set of CA3 pyramidal neurons that are held to overlap with the memory to be recalled as well as the firing of a set of pyramidal neurons not in the memory to be recalled; the firing of both sets of neurons is probably induced by synapses formed on CA3 neurons by perforant pathway axons. The firing of different sets of pyramidal neurons then evolves by discrete synchronous steps (for details see Bennett, Gibson, and Robinson 1994).

glial cells and neurons, it will be difficult to show whether the former are relevant or not when searching for cellular correlates, even though we are able to genetically manipulate the proteins in these different cell types. I think that it would be much wiser to consider the search for the "cellular" rather than just "neuronal" correlates. Furthermore we are at the beginning of understanding the variety of synaptic relationships between neurons, between glial cells, and between both classes of cells as well as the range of specialized synaptic mechanisms to be found amongst all of these cells. I believe that the present hubris in the neurosciences in which, for example, networks like those in figure 2 are taken as providing major insights into the workings of synaptic networks in any part of the brain is misplaced. In order to provide evidence to sustain this point I provide below some examples of the painfully slow progress being made in understanding even relatively simple synaptic networks that operate in the retina, the primary visual cortex, and the cerebellum.

NETWORKS IN THE RETINA

The simplest and most accessible part of the central nervous system is the retina, which during development appears first as an out-pocketing of the brain. For this reason the greatest histologist of the nervous system, Ramon y Cajal, considered the retina to be the ideal place at which to start research concerned with understanding the workings of synaptic networks in the central nervous system. Shortly after I began research, Barlow and Levick (1965) discovered what are called directionally selective ganglion neurons in the retina of some species. A directionally selective ganglion neuron is one that fires impulses at a high rate when an object is moved in one direction (therefore called the preferred direction) past the overlying light—sensitive rod photoreceptors that are connected to the ganglion cell. When the object moves in the opposite direction the ganglion neuron does not fire (and so this is called the null direction). Barlow and

Levick proposed the scheme shown in figure 3 (lower panel) to explain the network origins of directional selectivity. In this network a spatially offset inhibitory signal vetoes the excitatory signal for movement in the null direction. A lateral interneuron carries an inhibitory signal in the null but not the preferred direction, while the excitatory signal acts locally (figure 3). Therefore, inhibition arrives prior to, and can interact with, excitation for movement in the null direction, but inhibition lags behind excitation for movement in the preferred direction. This algorithm is spelt out in the legend to figure 3. I was very impressed by this simple network when it was published in 1967. This was the first analysis of its kind to offer an explanation for the operation of a real synaptic network.

If this algorithm is correct then the question arises as to the identity of the lateral interneuron and its synaptic connections. How far have we neuroscientists progressed in the past forty years since the algorithm was suggested in identifying the cellular components and their connections that carry out the necessary computations ? At the time of the original research on directional selectivity there were about ten different cell types recognized as composing the vertebrate retina (Ramon y Cajal 1904; Polyak 1941). Now at least fifty different cell types are recognized (Masland 2001), and this does not include the different types of glial cells that come into intimate contact with the neurons but do not conduct action potentials. It has taken four decades of research to identify some of the cellular mechanisms in the retina that are responsible for directional selectivity, with a number of important questions remaining to be answered (Fried, Munch, and Werblin 2005). A neuron having some of the characteristics specified in the Barlow and Levick (1965) scheme has been identified as the so-called starburst amacrine cell (Fried, Munch, and Werblin 2002). However the scheme has had to be radically modified with the discovery that the inputs to the ganglion cells are themselves directionally selective (Vaney and Taylor 2002). It appears then that there are several

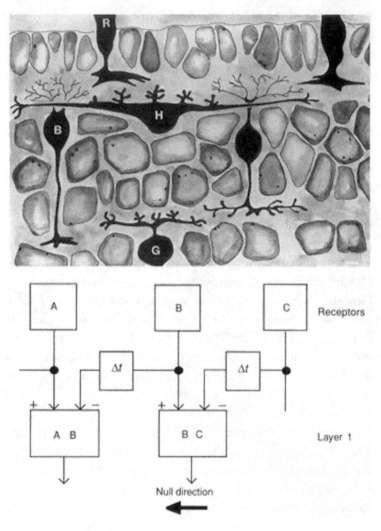

FIGURE 3. The upper panel shows a diagram of some of the principal cells in the retina as identified in 1965. R, rod photoreceptor; B, bipolar neuron; H, horizontal neuron; G, ganglion neuron connecting the retina to the brain.

The lower panel shows the algorithm proposed by Barlow and Levick (1965) to explain directional selectivity of ganglion neurons. A, B, and C are receptors that can respond to the object, which moves over them in either the null direction (indicated by the arrow) or in the preferred direction. These receptors can each excite activity in the units immediately beneath them. Each box containing a Δt is

(continued)

levels in the retinal circuitry that determine directional selectivity of ganglion neurons, which probably involve the activity of at least four different and specific retinal synaptic networks yet to be elucidated (Fried, Munch, and Werblin 2005). This illustrates the difficulties there are in understanding a network property of even what has been taken to be the simplest and most accessible part of the central nervous system.

NETWORKS IN THE PRIMARY VISUAL CORTEX

The most exciting scientific conference I have attended was over thirty years ago at the 1975 Cold Spring Harbor Symposium on the Synapse. I was invited by James Watson to present our findings on the plasticity of synapses between nerves and muscle cells, especially on our discoveries concerning the development of the initial point of contact of nerve terminal and muscle, how this subsequently becomes the site of excess numbers of nerve terminals followed by the elimination of all but one of these terminals as development proceeds (Bennett and Pettigrew 1976). Unbeknown to myself before arriving at the symposium, my presentation was to be followed by one from Hubel, Wiesel, and LeVay (1976). They gave a beautiful account of the development of synaptic network connections in the primary visual cortex (V1) that underlie the formation of columns of neurons that are dominated by connections with one eye or the other. Early during development these neurons have connections with both eyes, but by syn-

---

a unit that, if excited by the receptor connected to it, will, after a delay of length Δt, prevent excitation of the adjacent unit in the null direction. When an object moves in the null direction, electrical activity from an excited receptor (say C) excites (+) a unit in the layer immediately beneath it while at the same time inhibiting (-) the next unit in the null direction; each receptor in turn, namely, C, B, and A, carries out this process as the object moves over them. The delay units (shown as Δt) determine that the inhibitory process stops the excitatory activity from A and B moving through these gates if motion is in the null direction, but reaches the gates too late to produce such inhibition if motion is in the preferred direction.

aptic elimination one eye or the other comes to dominate in the connectivity. In a rather spectacular way they showed that there is considerable plasticity in this process, for if vision is confined to one eye during early development the other eye dominates the synaptic connectivity. This can be reversed during a critical period of development if vision is restored to both eyes.

Fifteen years after this memorable presentation Wiesel and his colleagues Gilbert and Hirsch returned to the topic of plasticity of synaptic connections in primary visual cortex (V1) at the 1990 Cold Spring Harbor Symposium. However this time the emphasis was on the extent of synaptic network plasticity in the adult visual cortex. Using electrophysiological recording techniques, their research indicated that there is large-scale reorganization of synaptic networks within the cortex a few months after a small retinal lesion (Gilbert, Hirsch, and Wiesel 1990). This work was taken to mean that adult cortical synaptic networks could readjust following loss of a sensory input signal (Gilbert 1998). Clearly understanding the plasticity mechanisms that are responsible for such readjustments is of considerable importance if we are to help those in need of appropriate remedial treatment.

At two to six months after binocular retinal lesions that deprive a zone within primary visual cortex (V1) of its normal input, stimulus-driven activity is reported to recur up to 5 mm inside the border of the cortex that has been deprived of its visual input (Gilbert 1998). More modest changes in cortical topography spanning 1–2 mm are thought to occur immediately (minutes to hours) after such lesions (Gilbert and Wiesel 1992). Fifteen years after this research on the plasticity of adult visual cortex, Logothetis and his colleagues studied signals in macaque primary visual cortex after small binocular retinal lesions in order to clarify the extent and time course of reorganization in the visual cortex (V1; Smirnakis et al. 2005). The retinal lesions were made by a photocoagulation laser and located in such a way as to create a homonymous visual field scotoma 4–8 degrees in diameter. These lesions deprived part of the cortex of visual input from each eye, which is a process thought to maximize reorganization. The region in the visual cortex that is deprived of retinal input is called the lesion projection zone. Functional magnetic resonance imaging was used to detect changes in the cortical topography of macaque visual cortex after these binocular retinal lesions. In contrast to the studies mentioned above using electrophysiology, the wide field of view provided by functional magnetic resonance imaging showed that primary visual cortex does not approach normal responsivity during 7.5 months following the retinal lesions, and its topography does not change (figure 4). Taking advantage of the wide field of view provided by imaging, electrophysiological recording electrodes could be placed precisely in the lesion projection zone. These confirmed the magnetic resonance imaging results. Thus, according to the observations of Logothetis and his colleagues, using two different techniques, primary visual cortex has limited potential for reorganization of synaptic networks , at least for several months following retinal injury.

The details of what might have gone awry in the previous fifteen years of research on the question of cortical plasticity have not yet been teased out. Suffice it to say that the complexity of determining the properties of this first site in the cortex that receives input from the retina is such as to require great care, technical skill, theoretical insight, and determination. Wiesel, a Nobel prize winner, together with Gilbert are neuroscientists of the first rank. Yet fifteen years of intensive research has not led to a consensus concerning the very important question of whether adult primary visual cortical neural networks are plastic (Giannikopoulos and Eysel 2006). I have spelt this story out not to apportion blame but to emphasize that the biological complexity of the synaptic network systems we are trying to understand is very considerable, testing the skills of even the best neuroscientists. Nevertheless, if we do not understand some of the fundamental properties of this first relay site in the cortex to other areas of cortex concerned with visual function, such as those in the temporal lobe whose normal function is required

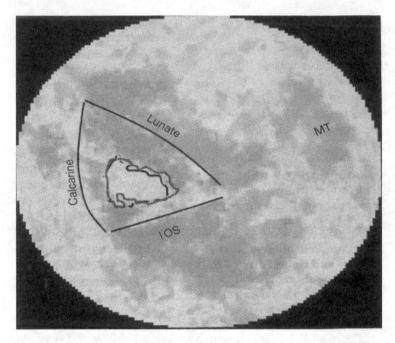

FIGURE 4. The blood oxygen level development (BOLD) signal, probably reflecting synaptic activity, measured by functional magnetic resonance imaging inside a lesion projection zone border does not change as a function of time after retinal lesioning. Shown is an area of visual cortex VI of radius 3 cm, centered near the foveal representation. The area is flattened and that outlined by the calcarine, lunate, and inferior occipital sulci is indicated. Regions outside this area largely correspond to nonvisual cortex. The lesion projection zone borders are shown at 0 days (inner contour), 4 months (outer contour), and 7.5 months (intermediate contour) after lesioning. The lesion projection zone was, respectively, 158, 179, and 180 mm² (from figure 2 in Smirnakis et al. 2005).

for face recognition, then we are very unlikely to have a firm grip on understanding the synaptic networks that subserve our visual capacities in such areas as the temporal lobe.

NETWORKS IN THE CEREBELLUM

The cerebellar cortex possesses a set of neuron types and synaptic connections that appear to be especially simple in arrangement and function, as shown in figure 5. This would appear to make the cerebellum especially amenable to experimental analysis as well as the development of empirically verifiable theories, a subject that I have found particularly fascinating (see, for example, Gibson, Robinson, and Bennett 1991). It was the brilliant David Marr (1969) who introduced the idea that acquisition of new motor skills depends on the plasticity of synapses between parallel axon terminals and Purkinje cell neurons in the cerebellum (figure 5). This appears to have been a tremendously fruitful suggestion for in the succeeding thirty-five years the plasticity of these synapses has been used to explain the acquisition and lifetime retention of many motor skills, including learned motor timing and reflex adaptation (Ito 2001). A large body of research has been carried out on the synaptic network model in which activity in climbing fiber axon terminals on Purkinje cells depress the strength of the parallel fiber axon terminal synapses on Purkinje cells when the two inputs (namely, climbing fiber axon terminals and parallel fiber axon terminals) are conjointly active (figure 5). This form of synaptic plasticity is said to underlie the acquisition of new motor skills. If electrical stimulation is used to directly stimulate climbing fiber axon terminals and parallel fiber axon terminals synchronously, then the amplitude of the synaptic potentials triggered by parallel fiber axon terminals in Purkinje cells is indeed depressed (Ito and Kano 1982). This depression of the synaptic potentials requires repeated pairings of climbing fiber axon terminal and parallel fiber axon terminal synaptic inputs. The depression is retained for many hours after the end of this protocol of stimulation and is called long-term depression. In the subsequent twenty-three years since the discovery of long-term depression by Ito and his colleagues there has been an immense amount of research on teasing out its molecular basis (Ito 2002).

Given the above, it was a great surprise last year when a critical experiment showed that long-term depression is not involved in motor learning in the cerebellum. Llinas and his colleagues

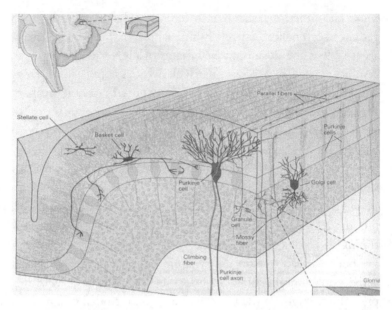

FIGURE 5. Cerebellar cortex. Shown are the large Purkinje cell neurons with their very large dendritic trees. These each receive synaptic connections from parallel fiber axon terminals derived from very small granule cell neurons that in turn receive synapses from mossy fiber axon terminals. The Purkinje cell neurons also, remarkably, receive synaptic connections from a single climbing fiber axon. This circuitry is laid out in a beautifully repeating and regular fashion, making the cerebellar cortex ideal for experimental investigation. Also shown are neurons of the inhibitory type (namely, Basket cells, Stellate cells, and Golgi cells).

(Welsh et al. 2005) prevented pharmacologically the long-term depression that arises at synapses formed by parallel fiber axon terminals on Purkinje cell neurons following conjunctive climbing fiber axon terminal and parallel fiber axon terminal stimulation (figure 6). This had no effect on the acquisition of motor skills involved in the rotorod test (Lalonde, Bensoula, and Filali 1995) nor in the development of motor timing during conditioning of the eyeblink reflex (McCormick and Thompson 1984). After thirty-six years of research, the synaptic networks and molecular mechanisms involved in cerebellar motor learning remain to be elucidated.

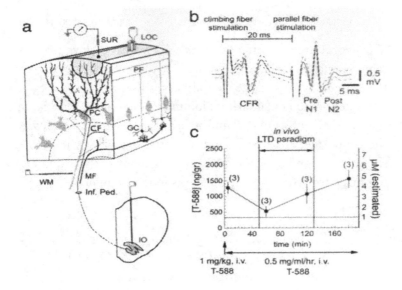

FIGURE 6. The drug T-588 prevents long-term depression of the parallel fiber to Purkinje cell synapse induced by conjunctive climbing fiber and parallel fiber stimulation in vivo. Similar concentrations of T-588 in the brains of behaving mice and rats affected neither motor learning in the rotorod test nor the learning of motor timing during classical conditioning of the eyeblink reflex. Thus parallel fiber to Purkinje cell long-term depression, under control of the climbing fibers, is not required for motor adaptation or the learning of response timing in two common models of motor learning for which the cerebellum has been implicated. In (a) are shown concentric bipolar stimulating electrodes placed onto the cerebellar surface (LOC) to stimulate a beam of parallel fibers (PFs) as well as placed into the cerebellar white matter (WM) to stimulate climbing fibers (CFs). A silver-ball surface electrode (SUR) is used to record evoked field potentials from Purkinje cell neurons (PCs). Direct stimulation of the inferior olive (IO) was used to verify that the potentials triggered by the WM electrode were climbing fiber responses (CFR). (b) shows CFR and presynaptic (N1) and postsynaptic (N2) PF responses triggered by conjunctive CF and PF stimulation using a 20 ms interstimulus interval to generate long-term depression for the N2 PF response. Dotted lines indicate one standard deviation. In (c) are shown the results for different T-588 concentrations in the brain. Four different time points are given during continuous intravenous infusion of T-588 with the long-term depression(LTD) paradigm performed between 50 and 130 min after infusion onset. The horizontal line at 1 uM indicates the concentration of T-588 that prevented LTD in vitro. The number of brains sampled at each time point is indicated in parentheses (from figure 1 in Welsh et al. 2005).

COMPLEXITY OF CENTRAL NERVOUS SYSTEM RESEARCH

The above three examples) illustrate the relatively slow progress neuroscience is making in teasing out the complexity of even the "simplest" parts of the central nervous system. They suggest that one should hesitate before joining in the hubris of thinking that neuroscientists understand many if any of the functions of the central nervous system and stress the extent to which one should pause and reflect before accepting many of the claims being made for what synaptic networks in the brain do.

Second Major Source of Concern Relating to Our Understanding of the Function of Cellular Networks in the Cortex

I mentioned two aspects of the engineering approach to understanding synaptic networks that give one pause for thought. I have considered the first of these in the second section of this chapter, namely, the great difficulty in teasing out biologically relevant properties pertinent for constructing an engineering type of network that may be used to further understanding of the biological network. The second difficulty arises when psychological attributes normally ascribed to humans and in some cases to other animals are attributed to synaptic networks, either before or after they have been reduced to engineering devices of varying degrees of complexity and modifiability. Particular synaptic networks or clusters of synaptic networks in the brain are said to remember, see, and hear. For example, it is suggested that "we can regard all seeing as a continual search for the answers to questions posed by the brain. The signals from the retina constitute 'messages' conveying these answers" (Young 1978). The visual cortex in the occipital pole (figure 7) is said to possess neurons that "present arguments on which the brain constructs its hypotheses of perception" (Blakemore 1977). As to the areas of

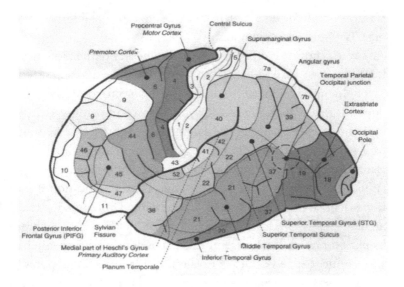

FIGURE 7. Cerebral cortex. Shown is a lateral view with the numbered areas designated by Brodmann (1909) as indicating cytologically distinguishable cellular classes and the relationships between them. Cellular networks concerned with vision are found, for example, in the occipital pole.

the brain that must function in order for us to see colors, these, it is claimed, are involved in "the interpretation that the brain gives to the physical property of objects (their reflectance), an interpretation that allows it to acquire knowledge rapidly about the property of reflectance" (Zeki 1999).

Indeed it is not just clusters of synaptic networks in the brain that are said to possess various psychological attributes but whole hemispheres of such networks (figure 7). For example, it is suggested that "the right hemisphere is capable of understanding language but not syntax" and that "the capacity of the right hemisphere to make inferences is extremely limited" (Gazzaniga, Ivry, and Mangun 2002). Further that "the left hemisphere can also perceive and recognize faces and can reveal superior capacities when the faces are familiar" and that "the left hemisphere adopts a helpful cognitive strategy in problem solving, but the right hemisphere does not have extra cognitive skills" (Gazza-niga, Ivry, and Mangun 2002).

These claims that synaptic networks, whether of the biological kind or after useful reduction to an engineering device, possess psychological attributes struck me as extraordinary. Although it does not follow logically, the slow and painstaking progress made by neuroscience in using the engineering approach to illumi-nate synaptic networks did not engender in me much hope that claims for their possessing psychological attributes could be sus-tained. I therefore sought help for conceptual clarification from those scholars that are professionally trained in such matters, namely, philosophers. After reading some of the major figures in philosophy of the last century, such as Russell, Wittgenstein, and Quine, I entered into discussion with some contemporary philosophers, in particular with Peter Hacker of Oxford. Our dialogue on the issue of whether psychological attributes might be ascribed to synaptic networks was carried out exclusively on the Internet and completed before we met. It has for me been an immensely satisfying journey. It has forced me to reconsider the history of neuroscience from Galen in the second century to the present time and to join with Peter in a critical analysis of the opinions of the giants of this discipline that have led neuroscien-tists into their present difficulties. This dialogue resulted in our book, *Philosophical Foundations of Neuroscience*. My satisfaction in this effort has been enhanced by the invitation of the American Philosophical Association to participate in a critical debate with Professors Dennett and Searle over the claims of our book. This has resulted in further clarification and in so doing enhanced our attempt to establish the truth concerning what the neurosciences have established and can hope to achieve in the future. In this way we have contributed to furthering the aims of neuroscience to assist in understanding what it means to be human and to ameliorate human suffering.

References

Barlow, H. B., and W. R. Levick. 1965. The mechanism of directionally selective units in rabbit's retina. *Journal of Physiology* 178:477–504.

Bennett, M. R. 2001. *History of the Synapse*. Australia: Harwood Academic.

Bennett, M. R., L. Farnell, and W. G. Gibson. 2005. A quantitative model of purinergic junctional transmission of calcium waves in astrocyte networks. *Biophysics Journal* 89:2235–50.

Bennett, M. R., W. G. Gibson, and J. Robinson. 1994. Dynamics of the CA3 pyramidal neuron autoassociative memory network in the hippocampus. *Philosophical Transactions of the Royal Society of London B Biological Sciences* 343:167–87.

Bennett, M. R., and P. M. S. Hacker. 2003. *Philosophical Foundations of Neuroscience*. Oxford: Blackwell.

Bennett, M. R., and A. G. Pettigrew. 1975. The formation of neuromuscular synapses. *Cold Spring Harbor Symposium in Quantitative Biology* 40:409–24.

Blakemore, C. 1997. *Mechanisms of the Mind*. Cambridge: Cambridge University Press.

Brindley, G. S. 1967. The classification of modifiable synapses and their use in models for conditioning. *Proceedings of the Royal Society London B* 168:361—76.

Brindley, G. S. 1969. Nerve net models of plausible size that perform many simple learning tasks. *Proceedings of the Royal Society London B* 174:173–91.

Brodman, K. 1909. *Vergleichende Lokalisation – lehre der Grosshirnrinde in ihren' Prinzipien dargestellt auf Grund des Zellenbaues*. Leipzig: Barth.

Cornell-Bell, A. H., S. M. Finkbeiner, M. S. Cooper, and S. J. Smith. 1990. Glutamate induces calcium waves in cultured astrocytes: long-range glial signalling. *Science* 247:470–73.

Fried, S. I., T. A. Munch, and F. S. Werblin. 2002. Mechanisms and circuitry underlying directional selectivity in the retina. *Nature* 420:411–14.

Fried, S. I., T. A. Munch, and F. S. Werblin. 2005. Directional selectivity is formed at multiple levels by laterally offset inhibition in the rabbit retina. *Neuron* 46:117–27.

Gazzaniga, M. S., R. B. Ivry, and G. R. Mangun. 2002. *Cognitive Neuroscience: The Biology of the Mind.* 2d ed. New York: Norton.

Giannikopoulos, D. V., and U. T. Eysel. 2006. Dynamics and specificity of cortical map reorganization after retinal lesions. *Proceedings of the National Academy of Sciences, USA* 103:1085–10.

Gibson, W. G., J. Robinson, and M. R. Bennett. 1991. Probabilistic secretion of quanta in the central nervous system: Granule cell synaptic control of pattern separation and activity regulation. *Philosophical Transactions of the Royal Society of London B Biological Sciences* 332:199–220.

Gilbert, C. D. 1998. Adult cortical dynamics. *Physiology Review* 78:467–85.

Gilbert, C. D., J. A. Hirsch, and T. N. Wiesel. 1990. Lateral interactions in visual cortex. *Cold Spring Harbor Symposium in Quantitative Biology* 55:663–77.

Gilbert, C. D., and T. N. Wiesel. 1992. Receptive field dynamics in adult primary visual cortex. *Nature* 356:150–52.

Hubel, D. H., T. N. Wiesel, and S. LeVay. 1976. Functional architecture of area 17 in normal and monocularly deprived macaque monkeys. *Cold Spring Harbor Symposium in Quantitative Biology* 40:581–89.

Ito, M. 2001. Cerebellar long-term depression: Characterization, signal transduction, and functional roles. *Physiology Review* 81:1143–95.

Ito, M. 2002. The molecular organization of cerebellar long-term depression. *Nature Reviews Neuroscience* 3:896–902.

Ito, M., and M. Kano. 1982. Long-lasting depression of parallel fiber—Purkinje cell transmission induced by conjunctive stimulation of parallel fibers and climbing fibers in the cerebellar cortex. *Neuroscience Letters* 33:253–58.

Koch, C. 2004. *The Quest for Consciousness.* CO: Roberts.

Lalonde, R., A. N. Bensoula, and N. Filali. 1995. Rotorod sensorimotor learning in cerebellar mutant mice. *Neuroscience Research* 22:423–26.

McCormick, D. A., and R. F. Thompson. 1984. Cerebellum: Essential involvement in the classically conditioned eyelid response. *Science* 223:296–99.

Marr, D. 1969. A theory of cerebellar cortex. *Journal of Physiology* 202:437–70.

Marr, D. 1971. Simple memory: A theory for archicortex. *Philosophical Transactions of the Royal Society of London B Biological Sciences* 262:23–81.

Masland, R. H. 2001. The fundamental plan of the retina. *Nature Neuroscience* 4:877–86.

Polyak, S. L. 1941. *The Retina*. Chicago: University of Chicago Press.

Ramon y Cajal, S. 1995 [1904]. *Histology of the Nervous System*. Trans. N. Swanson and L. W. Swanson. Oxford: Oxford University Press.

Sieburth, D., O. Ch'ng, M. Dybbs, M. Tavazoie, S. Kennedy, D. Wang, D. Dupuy, J. F. Rual, D. E. Hill, M. Vidal, G. Ruvkun, and J. M. Kapjan. 2005. Systematic analysis of genes required for synapse structure and function. *Nature* 436:510–17.

Smirnakis, S. M., A. A. Brewer, M. C. Schmid, A. S. Tolias, A. Schuz, M. Augath, W. Inhoffen, B. A. Wandell, and N. K. Logothetis. 2005. Lack of long-term cortical reorganization after macaque retinal lesions. *Nature* 435:300–7.

Vaney, D. I., W. R. Taylor. 2002. Direction selectivity in the retina. *Current Opinion in Neurobiology* 12:405–10.

Welsh, J. P., H. Yamaguchi, X. H. Zeng, M. Kojo, Y. Nakada, A. Takagi, M. Sugimori, and R. R. Llinas. 2005. Normal learning during pharmacological prevention of Purkinje cell long-term depression. *Proceedings of the National Academy of Sciences, USA* 102:17166–71.

Young, J. Z. 1978. *Programs of the Brain*. Oxford: Oxford University Press.

Zeki, S. 1999. Splendours and miseries of the brain. *Philosophical Transactions of the Royal Society of London B Biological Sciences* 354:2053–65.

THE REBUTTALS

PHILOSOPHY AS NAIVE ANTHROPOLOGY

Comment on Bennett and Hacker

DANIEL DENNETT

Bennett and Hacker's *Philosophical Foundations of Neuroscience* (Blackwell, 2003), a collaboration between a philosopher (Hacker) and a neuroscientist (Bennett), is an ambitious attempt to reformulate the research agenda of cognitive neuroscience by demonstrating that cognitive scientists and other theorists, myself among them, have been bewitching one another by misusing language in a systematically "incoherent" and conceptually "confused" way. In both style and substance, the book harks back to Oxford in the early 1960s, when Ordinary Language Philosophy ruled and Ryle and Wittgenstein were the authorities on the meanings of our everyday mentalistic or psychological terms. I myself am a product of that time and place (as is Searle, for that matter), and I find much to agree with in their goals and presuppositions and, before turning to my criticisms, which will be severe, I want to highlight what I think is exactly right in their approach—the oft-forgotten lessons of Ordinary Language Philosophy.

Neuroscientific research ... abuts the psychological, and clarity regarding the achievements of brain research presupposes clarity regarding the categories of ordinary psychological description—that is, the categories of sensation and perception, cognition and recollection, cogitation and imagination, emotion and volition. To the extent that neuroscientists fail to grasp the contour lines of the relevant categories, they run the risk not only of asking the wrong questions, but also of misinterpreting their own experimental results.   (p. 115)

Just so.[1] When neuroscientists help themselves to the ordinary terms that compose the lore I have dubbed "folk psychology,"[2] they need to proceed with the utmost caution, since these terms have presuppositions of use that can subvert their purposes and turn otherwise promising empirical theories and models into thinly disguised nonsense. A philosopher—an expert on nuances of meaning that can beguile the theorist's imagination—is just the right sort of thinker to conduct this important exercise in conceptual hygiene.

I also agree with them (though I would not put it their way) that "the evidential grounds for the ascription of psychological attributes to others are not inductive, but rather criterial; the evidence is logically good evidence" (p. 82). This puts me on their side against, say, Fodor.[3]

So I agree wholeheartedly with the motivating assumption of their book. I also applaud some of their main themes of criticism, in particular their claim that there are unacknowledged Cartesian leftovers strewn everywhere in cognitive neuroscience and causing substantial mischief. They say, for instance:

Contemporary neuroscientists by and large take colours, sounds, smells and tastes to be "mental constructions created in the brain by sensory processing. They do not exist, as such, outside the brain" [quoting Kandel et al. 1995].

This again differs from Cartesianism only in replacing the mind by the brain.   (p. 113)

Here they are criticizing an instance of what I have called "Cartesian materialism" (*Consciousness Explained*, 1991), and they are right, in my opinion, to see many cognitive neuroscientists as bedazzled by the idea of a place in the brain (which I call the Cartesian Theater) where an inner show of remarkable constructions is put on parade for a (material) *res cogitans* sitting in the audience.

More particularly, I think they are right to find crippling Cartesianism in Benjamin Libet's view of intentional action and in some of the theoretical work by Stephen Kosslyn on mental imagery. I also join them in deploring the philosopher's "technical" term, *qualia*, a poisoned gift to neuroscience if ever there was one, and I share some of their misgivings about the notorious "what is it like" idiom first explored by Brian Farrell (1950) and made famous by Thomas Nagel (1974). Introspection, they say, is not a form of inner vision; there is no mind's eye. I agree. And when you have a pain, it isn't like having a penny; the pain *isn't a thing* that is in there. Indeed. Although I don't agree with everything they say along the paths by which they arrive at all these destinations, I do agree with their conclusions. Or, more accurately, they agree with my conclusions, though they do not mention them.[4]

More surprising to me than their failure to acknowledge these fairly substantial points of agreement is that the core of their book, which is also the core of their quite remarkably insulting attack on me,[5] is a point I myself initiated and made quite a big deal of back in 1969. Here is what they call the mereological fallacy:

We know what it is for human beings to experience things, to see things, to know, or believe things, to make decisions, to interpret equivocal data, to guess and to form hypotheses.

But do we know what it is for a *brain* to see or hear, for a *brain* to have experiences, to know or believe something? Do we have any conception of what it would be for a *brain* to make a decision?

They answer with a ringing NO!

It makes no sense to ascribe psychological predicates (or their negations) to the brain, save metaphorically or metonymically. The resultant combination of words does not say something that is false; rather, it says nothing at all, for it lacks sense. Psychological predicates are predicates that apply essentially to the whole living animal, not to its parts. It is not the eye (let alone the brain) that sees, but *we* see *with* our eyes (and we do not see with our brains, although without a brain functioning normally in respect of the visual system, we would not see).   (p. 72)

This is at least close kin to the point I made in 1969 when I distinguished the personal and subpersonal levels of explanation. *I* feel pain; my brain doesn't. *I* see things; my eyes don't. Speaking about pain, for instance, I noted:

An analysis of our ordinary way of speaking about pains shows that no events or processes could be discovered in the brain that would exhibit the characteristics of the putative "mental phenomena" of pain, because talk of pains is essentially non-mechanical, and the events and processes of the brain are essentially mechanical.

(*Content and Consciousness,* p. 91)

We have so much in common, and yet Bennett and Hacker are utterly dismissive of my work. How can this be explained? As so often in philosophy, it helps to have someone say, resolutely and clearly, what others only hint at or tacitly presuppose. Bennett and

Hacker manage to express positions that I have been combating *indirectly* for forty years but have never before been able to confront head on, for lack of a forthright exponent. Like Jerry Fodor, on whom I have relied for years to blurt out vividly *just* the points I wish to deny—saving me from attacking a straw man—Bennett and Hacker give me a bold doctrine to criticize. I've found the task of marshaling my thoughts on these topics in reaction to their claims to be illuminating to me and, I hope, to others as well.

The Philosophical Background

In this section I am going to speak just of Hacker, leaving his coauthor Bennett out of the discussion, since the points I will be criticizing are clearly Hacker's contribution. They echo, often in the same words, claims he made in his book, *Wittgenstein: Meaning and Mind* (Blackwell, 1990), and they are strictly philosophical.

When Hacker lambastes me, over and over, for failing to appreciate the mereological fallacy, this is a case of teaching your grandmother to suck eggs. I am familiar with the point, having pioneered its use. Did I, perhaps, lose my way when I left Oxford? Among the philosophers who have taken my personal level/subpersonal level distinction to heart, at least one—Jennifer Hornsby—has surmised that I might have abandoned it in my later work.[6] Did I in fact turn my back on this good idea? No.[7] On this occasion it would be most apt to cite my 1980 criticism of Searle's defense of the Chinese Room intuition pump:

> The systems reply suggests, entirely correctly in my opinion, that Searle has confused different levels of explanation (and attribution). *I* understand English, my brain doesn't—nor, more particularly, does the proper part of it (if such can be isolated) that operates to "process" incoming sentences and to execute my speech act intentions.
>
> (*Behavioral and Brain Sciences* [1980], 3:429)[8]

(This claim of mine was summarily dismissed by Searle, by the way, in his reply in BBS. I'll be interested to see what he makes of the personal level/subpersonal level distinction in its guise as the mereological fallacy.)[9]

The authoritative text on which Hacker hangs his conviction about the mereological fallacy is a single sentence from St. Ludwig:

> It comes to this: Only of a human being and what resembles (behaves like) a living human being can one say: it has sensations; it sees, is blind; hears, is deaf; is conscious or unconscious.   (*Philosophical Investigations*, para. 281)

Right here is where Hacker and I part company. I am happy to cite this passage from Wittgenstein myself; indeed I take myself to be *extending* Wittgenstein's position: I see that robots and chess-playing computers and, yes, brains and their parts *do* "resemble a living human being (by behaving like a human being)"—and this resemblance is sufficient to warrant an adjusted use of psychological vocabulary to characterize that behavior. Hacker does not see this, and he and Bennett call all instances of such usage "incoherent," insisting again and again that they "do not make sense." Now who's right?

Let's go back to 1969 and see how I put the matter then:

> In one respect the distinction between the personal and sub-personal levels of explanation is not new at all. The philosophy of mind initiated by Ryle and Wittgenstein is in large measure an analysis of the concepts we use at the personal level, and the lesson to be learned from Ryle's attacks on "para-mechanical hypotheses" and Wittgenstein's often startling insistence that explanations come to an end rather earlier than we had thought is that the personal and sub-personal levels must not be confused. The lesson has occasionally been misconstrued, however, as the lesson that

the personal level of explanation is the only level of
explanation when the subject matter is human minds
and actions. In an important but narrow sense this is true,
for as we see in the case of pain, to abandon the personal
level is to stop talking about pain. In another important
sense it is false, and it is this that is often missed. The recognition that there are two levels of explanation gives birth to
the burden of relating them, and this is a task that is not
outside the philosopher's province. . . . There remains the
question of how each bit of the *talk* about pain is related
to neural impulses or talk about neural impulses. This and
parallel questions about other phenomena need detailed
answers even after it is agreed that there are different sorts
of explanation, different levels and categories.

(*Content and Consciousness,* pp. 95–96)

This passage outlines the task I have set myself during the
last thirty-five years. And the boldfaced passages mark the main
points of disagreement with Hacker, for my path is not at all the
path that he has taken. He gives his reasons, and they are worth
careful attention:

[A] Conceptual questions antecede matters of truth and
falsehood. . . . Hence conceptual questions are not amenable to scientific investigation and experimentation or to
scientific theorizing.   (p. 2)

One can wonder about the first claim. Are not answers to these conceptual questions either true or false? No, according to Hacker:

[B] What truth and falsity is to science, sense and nonsense
is to philosophy.   (p. 6)

So when philosophers make mistakes they produce nonsense,
never falsehoods, and when philosophers do a good job we

mustn't say they get it *right* or speak the *truth* but just that they make sense.[10] I am inclined to think that Hacker's [B] is just plain *false*, not nonsense, but, be that as it may, Hacker's second claim in [A], in spite of the "hence," is a non sequitur. Even if conceptual questions do "antecede" matters of truth and falsity, it might well behoove anybody who wanted to get clear about what the good answers are to investigate the relevant scientific inquiries assiduously. This proposal, which Hacker identifies as Quinian naturalism, he dismisses with an irrelevancy: "we do not think that empirical research can solve any philosophical problems, any more than it can solve problems in mathematics" (p. 414). Well of course not; empirical research doesn't *solve* them, it *informs* them and sometimes *adjusts* or *revises* them, and then they sometimes *dissolve*, and sometimes they can then be solved by further philosophical reflection.

Hacker's insistence that philosophy is an a priori discipline that has no continuity with empirical science is the chief source of the problems bedeviling this project, as we shall see:

[C] How can one investigate the bounds of sense? Only by examining the use of words. Nonsense is often generated when an expression is used contrary to the rules for its use. The expression in question may be an ordinary, non-technical expression, in which case the rules for its use can be elicited from its standard employment and received explanations of its meaning. Or it may be a technical term of art, in which case the rules for its use must be elicited from the theorist's introduction of the term and the explanations he offers of its stipulated use. Both kinds of terms can be misused, and when they are, nonsense ensues—a form of words that is excluded from the language. For either nothing has been stipulated as to what the term means in the aberrant context in question, or this form of words is actually excluded by a rule specifying that there is no such thing as (e.g., that there is no such thing as

"east of the North Pole"), that this is a form of words that
has no use. (p. 6)

This passage is all very reminiscent of 1960 or thereabouts, and
I want to remind you of some of the problems with it, which I
had thought we had figured out many years ago—but then, we
didn't have this forthright version to use as our target.

How can one investigate the bounds of sense? Only by
examining the use of words.

Notice, first, that, no matter what any philosopher may say, examining the use of words is an empirical investigation, which
often yields everyday garden-variety truths and falsehoods and
is subject to correction by standard observations and objections.
Perhaps it was a dim appreciation of this looming contradiction
that led Hacker, in his 1990 book, to pronounce as follows:

Grammar is autonomous, not answerable to, but presupposed by, factual propositions. In this sense, unlike means/
ends rules, it is arbitrary. But it has a kinship to the non-
arbitrary. It is moulded by human nature and the nature of
the world around us. (p. 148)

Let grammar be autonomous, whatever that means. One still
cannot study it without asking questions—and even if you only
ask *yourself* the questions, you still have to see what you say. The
conviction that this method of consulting one's (grammatical or
other) intuitions is entirely distinct from empirical inquiry has a
long pedigree (going back not just to the Oxford of the 1960s,
but to Socrates), but it does not survive reflection.

This can be readily seen if we compare this style of philosophy with anthropology, a manifestly empirical inquiry that can
be done well or ill. If one chooses second-rate informants, or
doesn't first get quite fluent in their language, one is apt to do

third-rate work, For this reason, some anthropologists prefer to do one or another form of *autoanthropology*, in which you use yourself as your informant—perhaps abetted by a few close colleagues as interlocutors. The empirical nature of the enterprise is just the same.[11] Linguists, famously, engage in a species of this autoanthropology, and they know a good deal, at this point, about the pitfalls and risks of their particular exercises in teasing out grammatical intuitions regarding their native tongues. It is well known, for instance, that it is very difficult to avoid contaminating your intuitions about grammaticality with your own pet theoretical ideas. Some linguists, in fact, have been led to the view that theoretical linguists are, or should be, disqualified as informants, since their judgments are not naive. Now here is a challenge for Hacker and like-minded philosophers: How, precisely, do they distinguish their inquiry from autoanthropology, an empirical investigation that apparently uses just the same methods and arrives at the same sorts of judgments.[12]

Anybody who thinks that philosophers have found a method of *grammatical* inquiry that is somehow immune to (or orthogonal to or that "antecedes") the problems that can arise for that anthropological inquiry owes us an apologia explaining just how the trick is turned. Bald assertions that *this is what philosophers do* only evade the challenge. My colleague Avner Baz reminds me that Stanley Cavell[13] has made an interesting move toward meeting this burden: Cavell claims that the philosopher's observations about what we would say are more akin to *aesthetic* judgments. As Baz puts it, "you present your judgment as *exemplary*—you talk *for* a community" (personal communication), and this is fine as far as it goes, but since the anthropologist is also engaged in finding *the best, most coherent* interpretation of the data gathered (Quine's point about the principle of charity, and my point about the rationality assumption of the intentional stance), this normative or commendatory element is already present—but bracketed—in the anthropologist's investigation. The anthropologist cannot make sense of what his informants *say* without uncovering what they *ought to say* (in their own community) under many conditions. What is deliberately left out of the anthropologist's enterprise, however, and what needs defense in the philosopher's enterprise, is a *justification* for the following claim: This is what these people do and say, *and you should do the same.* As we shall see, it is Hacker's failure to identify the community he is speaking for that scuttles his project.

Back to [C]:

Nonsense is often generated when an expression is used contrary to the rules for its use.

It is long past time to call a halt to this sort of philosophical pretense. Ryle notoriously claimed to identify "category mistakes" by appeal to the "logic" of existence claims, but, let's face it: that was a bluff. He had no articulated logic of existence terms to back up his claims. In spite of the popularity of such talk, from Ryle and Wittgenstein and a host of imitators, no philosopher has *ever* articulated "the rules" for the use of any ordinary expression. To be sure, philosophers have elicited judgments of deviance by the hundreds, but noting that "we wouldn't say thus-and-so" is not expressing a rule. Linguists use an asterisk or star to make the same sort of point, and they have generated thousands of starred sentences such as

*An acorn grew into every oak.
*The house was rats infested.

But, as any linguist will assure you, drawing attention to a judgment of deviance—even if it is part of a large and well-described pattern of deviance—is not the same as uncovering the rules that govern those cases. Linguists have worked very hard for over forty years to articulate the rules of English syntax and semantics and have a few modest corners in which they can plausibly claim to have elicited "the rules." But they also have encountered large areas of fuzziness. What about this sentence?

*The cat climbed down the tree [an example from Jack-
endoff].

Is this nonsense that violates "the rules" of the verb *to climb*? It's
hard to say, and it may be that usage is changing. Such examples
abound. Linguists have learned that something may sound a
bit odd, smell a bit fishy, but still not violate any clear rule that
anybody has been able to compose and defend. And the idea of
rules that are ineffable is too obscurantist to be worth discussion. Philosophers' intuitions, no matter how sharply honed,
are not a superior source of evidence in this manifestly empirical inquiry.

Back to [C]. Hacker goes on to divide the lexicon in two:

> The expression in question may be an ordinary, non-tech-
> nical expression, in which case the rules for its use can be
> elicited from its standard employment and received expla-
> nations of its meaning. Or it may be a technical term of art,
> in which case the rules for its use must be elicited from the
> theorist's introduction of the term and the explanations he
> offers of its stipulated use.

I am tempted to assert that Hacker is just wrong (but not
speaking nonsense) when he implies that the hallmark of a
technical term is that it is "introduced" by a theorist who "stip-
ulates" its use. Either that, or he is defining "technical term" so
narrowly that many of the terms we would ordinarily agree to
be technical wouldn't be so classified by him—and "technical
term" is a technical term whose use he is stipulating here and
now. Let Hacker have his definition of technical terms, then,
narrow though it is. None of the terms that are the focus of the
attacks in the book are technical in this sense, so they must be
"ordinary, non-technical" terms—or they must be mongrels,
a possibility that Hacker briefly considers in his 1990 book
and dismisses:

If neurophysiologists, psychologists, artificial-intelligence scientists, or philosophers wish to change existing grammar, to introduce new ways of speaking, they may do so; but their new stipulations must be explained and conditions of application laid down. What may not be done is to argue that since we know what "to think," "to see," or "to infer" mean and know what "the brain" means, therefore we must know what "the brain thinks, sees, and infers" means. For we know what these verbs mean only in so far as we have mastered their existing use, which does not license applying them to the body or its parts, save derivatively. Nor may one cross the new "technical" use with the old one, as, for example, neuroscientist typically do in their theorizing. For this produces a conflict of rules and hence incoherence *in the neuroscientists' use* of these terms.

(pp. 148–49).

This last claim—and it is also at the very heart of the 2003 book—is question begging. If Hacker were able to *show us* the rules, and show us just how the new uses conflict with them, we might be in a position to agree or disagree with him, but he is just making this up. He has no idea what "the rules" for the use of these everyday psychological terms are. More tellingly, his insistence on an a prioristic methodology systematically blinds him to what he is doing here. Let him be *right*[14] in his conviction that he has an a priori method that gives him "antecedent" insight into the meanings of his ordinary psychological terms. He still needs to confront the burden of showing how his prolegomenon or stage setting avoids the pitfall of what we might call conceptual myopia: treating *one's own* (possibly narrow and ill-informed) concepts as binding on others with different agendas and training. How, indeed, does he establish that he and those whose work he is criticizing are speaking the same language? That is surely an empirical question, and his failure to address it with sufficient care has led him astray. What he has done, in fact, is not good philosophy but bad anthropology: he went to cognitive science to "examine the use of words" and failed to notice that he himself was bringing *his* ordinary language into alien territory, and that *his* intuitions didn't necessarily apply. When he calls *their* usage "aberrant," he is making a beginner's mistake.

The use of psychological predicates in the theorizing of cognitive scientists is indeed a particular patois of English, quite unlike the way of speaking of Oxford philosophy dons, and it has its own "rules." How do I know this? Because I've done the anthropology. (You have to be a Quinian naturalist to avoid making these simple mistakes.) There is a telling passage in which Hacker recognizes this possibility but exposes his inability to take it seriously:

> Is it a new discovery that brains also engage in such human activities? Or is it a linguistic innovation, introduced by neuroscientists, psychologists and cognitive scientists, extending the ordinary use of these psychological expressions for good theoretical reasons? Or, more ominously, is it a conceptual confusion?   (pp. 70–71)

Hacker opts for the third possibility, without argument, while I say it's the first two together. There *is* an element of discovery. It is an empirical fact, and a surprising one, that our brains—more particularly, *parts* of our brains—engage in processes that are *strikingly like* guessing, deciding, believing, jumping to conclusions, etc. And it is *enough* like these personal level behaviors to warrant stretching ordinary usage to cover it. If you don't study the excellent scientific work that this adoption of the intentional stance has accomplished, you'll think it's just crazy to talk this way. It isn't.

In fact this is what inspired me to develop my account of the intentional stance. When I began to spend my time talking with researchers in computer science and cognitive neuroscience, what struck me was that they unself-consciously, without any nudges or raised eyebrows, spoke of computers (and programs and subroutines and brain parts and so forth) *wanting* and *thinking* and *concluding* and *deciding* and so forth. What are the rules? I asked myself. And the answer I came up with are the rules for adopting the intentional stance. The *factual* question is: do people in these fields speak this way, and does the intentional stance capture at least a central part of "the rules" for how they speak? And the (factual) answer is Yes.[15] There is also, I suppose, a political question: Do they have any right to speak this way? Well, it pays off handsomely, generating hypotheses to test, articulating theories, analyzing distressingly complex phenomena into their more comprehensible parts, and so forth.

Hacker also discovers this ubiquitous use of intentional terms in neuroscience, and he's shocked, I tell you, shocked! So many people making such egregious conceptual blunders! He doesn't know the half of it. It is not just neuroscientists; it is computer scientists (and not just in AI), cognitive ethologists, cell biologists, evolutionary theorists ... all blithely falling in with the game, teaching their students to think and talk this way, a linguistic pandemic. If you asked the average electrical engineer to explain how half the electronic gadgets in your house worked, you'd get an answer bristling with intentional terms that commit the mereological fallacy—if it is a fallacy.

It is not a fallacy. We don't attribute *fully fledged* belief (or decision or desire—or pain, heaven knows) to the brain parts— that *would* be a fallacy. No, we attribute an attenuated sort of belief and desire to these parts, belief and desire stripped of many of their everyday connotations (about responsibility and comprehension, for instance). Just as a young child can *sort of* believe that her daddy is a doctor (without full comprehension of what a daddy or a doctor is),[16] so a robot—or some part of a person's brain—can *sort of* believe that there is an open door a few feet ahead, or that something is amiss over there to the right, and so forth. For years I have defended such uses of the intentional stance in characterizing complex systems ranging from chess-playing computers to thermostats and in characterizing the brain's subsystems at many levels. The idea is that, when we engineer a complex system (or reverse engineer a biological system like a person or a person's brain), we can make progress by breaking down the whole wonderful person into subpersons of sorts agentlike systems that have *part* of the prowess of a person, and then these homunculi can be broken down further into still simpler, less personlike agents, and so forth—a *finite*, not infinite, regress that bottoms out when we reach agents so stupid that they can be replaced by a machine. Now perhaps all my attempts at justifying and explaining this move are mistaken, but since Bennett and Hacker never address them, they are in no position to assess them.

Here is how I put it in a paper Hacker cites several times (though not this passage):

> One may be tempted to ask: Are the subpersonal components *real* intentional systems? At what point in the diminution of prowess as we descend to simple neurons does *real* intentionality disappear? Don't ask. The reasons for regarding an individual neuron (or a thermostat) as an intentional system are unimpressive, but not zero, and the security of our intentional attributions at the highest levels does not depend on our identifying a lowest level of real intentionality.
>
> ("Self-portrait," in Guttenplan, listed by H & B as
> "Dennett, Daniel C. Dennett," 1994)

The homunculus *fallacy*, by attributing the *whole* mind to a proper part of the system, merely postpones analysis and thus would generate an infinite regress since each postulation would make no progress. Far from it being a *mistake* to attribute hemi-semi-demi-proto-quasi-pseudo intentionality to the mereological parts of persons, it is precisely the enabling move that lets us see how on earth to get whole wonderful persons out of brute mechanical parts. That is a devilishly hard thing to imagine, and the poetic license granted by the intentional stance eases the task substantially.[17] From my vantage point, then, Hacker is comically naive, for all the world like an old-fashioned grammarian scolding people for saying "ain't" and insisting *you can't say that!* to people who manifestly *can* say that and know what they mean when they do. Hacker had foreseen this prospect in his 1990 book and described it really very well:

> If all this [cognitive science] is to be taken at face value, it seems to show, first, that the grammatical remark that these predicates, in their literal use, are restricted to human beings and what behaves like human beings is either wrong *simpliciter* or displays "semantic inertia" that has been overtaken by the march of science, for machines actually do behave like human beings. Secondly, if it makes literal sense to attribute epistemic and even perceptual predicates to machines which are built to simulate certain human operations and to execute certain human tasks, it seems plausible to suppose that the human brain must have a similar abstract functional structure to that of the machine design. In which case, surely it must make sense to attribute the variety of psychological predicates to the human brain after all. (pp. 160–61)

Exactly. That's the claim. How does he rebut it? He doesn't. He says, "Philosophical problems stem from conceptual confusion. They are not resolved by empirical discoveries, and they cannot be answered, but only swept under the carpet, by conceptual change" (p. 161). Since Hacker's philosophical problems are becoming obsolete, I suppose we might just sweep them under the carpet, though I'd prefer to give them a proper burial.

The Neuroscientific Details

When Bennett and Hacker undertake to examine the neuroscientific literature, there is scant variety to their critique. They quote Crick and Edelman and Damasio and Gregory and many others saying things that strike them as "incoherent" because these scientists commit the so-called mereological fallacy.

> Far from being new homonyms, the psychological expressions they use are being invoked in their customary sense, otherwise the neuroscientists would not draw the inferences from them which they do draw. When Crick asserts that "what you see is not what is *really* there; it is what your brain *believes* is there," it is important that he takes "believes" to have its normal connotations—that it does not mean the same as some novel term "believes.*" For it is part of Crick's tale that the belief is the outcome of an *interpretation* based on previous *experience* or *information* (and not the outcome of an interpretation* based on previous experience* and information.*
>
> (p. 75)

In fact they are just wrong (but not nonsensical). Crick's whole tale (and in this instance it is a quite banal and uncontroversial explanation) is intended by Crick to be understood at the sub-personal level. The interpretation in question is *not* of (personal level) *experience* but of, say, *data from the ventral stream*, and the process of interpretation is of course supposed to be a subpersonal process. Another passage in the same spirit:

> Similarly, when [J. Z.] Young talks of the brain's containing knowledge and information which is encoded in the brain "just as knowledge can be recorded in books or computers," he means knowledge, not knowledge*—since it is knowledge and information, not knowledge* and information*, that can be recorded in books and computers.

The authors have done nothing at all to establish that there is no concept of knowledge or information that can be encoded in both books and brains. There is a large and complex literature in cognitive science on the concept of information—and on the concept of knowledge (just think of Chomsky's discussions of "cognizing" in response to an earlier parry in much the same direction as the authors')—and the authors' obliviousness to these earlier discussions shows that they are not taking their task very seriously. Many other examples in the same vein could be cited. They have one idea, the mereological fallacy, and they use it wholesale, without any consideration of the details. Each time they quote the offending passage—and they could have found a hundred times more instances of intentional stance attributions to brain subsystems—and then simply declare it nonsense because it commits their fallacy. Not once do they attempt to show that because of making this presumably terrible mistake the author in question is led astray into some actual error or contradiction. Who knew philosophy of neuroscience would be so easy?

Consider their discussion of the fascinating and controversial topic of mental imagery. First they demonstrate—but I doubt anybody has ever doubted it—that creative imagination and mental imagery are really quite distinct and independent phenomena. Then comes their knockout punch: "A topographically arranged sensory area is not an image of anything; there are no images in the brain, and the brain does not *have* images" (p. 183). As one who has argued strenuously for years that we must not jump to the conclusion that "mental imagery" involves actual images in the brain, and that the retinotopic arrays found therein may well not *function* as images in the brain's processing, I must note that their bald assertion does not help at all. It is simply irrelevant whether "we would say" that the brain *has images*. Whether any of the arrays of stimulation in the brain that manifestly have the geometric properties of images *function* as images is an empirical question, and one that is close to being answered.

Philosophical analysis is powerless to settle the issue—except by a deeply reactionary insistence that these imagelike data structures, from which information is apparently extracted in much the way we persons (at the personal level) extract information visually from public images, don't *count* as images. Such obfuscatory moves give philosophy a serious credibility problem in cognitive science.

In fact, there are serious conceptual problems with the ways in which cognitive scientists have spoken about images and knowledge and representations and information and the rest. But it is hard, detailed work *showing* that the terminology used is being misused in ways that seriously mislead the theorists. The fact is that, for the most part, these terms, as they are found in cognitive science, really are "ordinary language"—not technical terms[18] that have been explicitly stipulated within some theory. Theorists have often found it useful to speak, *somewhat* impressionistically, about the information being processed, the decisions being reached, the representations being consulted, and, instead of doing what a philosopher might do when challenged about what they meant, namely, *defining their terms more exactly*, they instead point to their models and say: "See: here are the mechanisms at work, doing the information processing I was telling you about." And the models work. They *behave* in the way they have to behave in order to cash out *that* particular homunculus, so there need be no further cavil about exactly what was being attributed to the system.

But there are also plenty of times when theorists' enthusiasm for their intentional interpretations of their models misleads them.[19] For instance, in the imagery debate, there have been missteps of overinterpretation—by Stephen Kosslyn, for instance—that need correction. It is not that map talk or image talk is *utterly* forlorn in neuroscience, but that it has to be very carefully introduced, and it sometimes isn't. Can philosophy help? Yes it can, say Bennett and Hacker: "It can explain—as we have explained—why mental images are not ethereal pictures and why they cannot be rotated in mental space" (p. 405). This wholesale approach is not helpful. What is actually happening in the brain when people are engaged in mental imagery cannot be settled by making the point that the personal level is not the subpersonal level. The theorists already know that; they are not making *that* mistake. They are actually quite careful and subtle thinkers, and some of them *still* want to talk about images functioning as images in the brain. They may well be right.[20] Philosophers such as Hacker may lose interest once the topic is subpersonal,[21] but then they shouldn't make the mistake of criticizing a domain they know little about.

Sometimes the authors miss the point with such blithe confidence that the effect is quite amusing, as in their stern chastising of David Marr:

> To see is not to discover anything from an image or light array falling upon the retina. For one cannot, in this sense, discover anything from something one cannot perceive (we do not perceive the light array that falls upon our retinae [*sic*], what we perceive is whatever that light array enables us to perceive. (p. 144)

Got it. Since *we* do not perceive the light array that falls upon our retinas, it is obvious that *we* do not discover anything. Marr was not an idiot. He understood that. Now, what about Marr's theory of the *subpersonal processes* of vision?

> Moreover, it is *altogether obscure* [emphasis added] how the mind's having access to putative neural *descriptions* will enable the person *to see*. And if Marr were to insist (rightly) that it is the person, not the mind, that sees, how is the transition from the presence of an encoded 3-D model description in the brain to the experience of seeing what is before one's eyes to be explained? To be sure, *that* is not an empirical problem, to be solved by further investigations. It

is the product of a conceptual confusion, and what it needs
is disentangling. (p. 147)

I would have said, on the contrary, that it is a philosophical problem to be solved by addressing those who find it "altogether obscure" and leading them to an understanding of just how Marr's
theory *can* account for the family of competences that a seer
has.[22] Marr was more or less taking it for granted that his readers
could work out for themselves how a model of the brain as having a consultable 3-D model of the world would be well on the
way to explaining how a creature with just such a brain could
see, but, if this eluded some readers, a philosopher would probably be a good specialist to explain the point. Just *asserting* that
Marr is suffering from a conceptual confusion has, as Russell so
aptly put it, all the advantages of theft over honest toil.

> For seeing something is the exercise of a power, a use of
> the visual faculty—*not* [emphasis added] the processing of
> information in the semantic sense or the production of a
> description in the brain. (p. 147)

This *not* is another theft. What has to be explained is the power
of the "visual faculty," and that power is explained in terms of the
lesser powers of its parts, whose activities include the creation
and consultation of descriptions (of sorts). These examples could
be multiplied to the point of tedium:

> It makes no sense, save as a misleading figure of speech, to
> say, as LeDoux does, that it is "possible for your brain to
> know that something is good or bad before it knows ex
> actly what it is." (p. 152)

But who is misled? Not LeDoux, and not LeDoux's readers, if
they read carefully, for they can see that he has actually found

a very good way to make the surprising point that a specialist circuit in the brain can discriminate something as dangerous, say, or as desirable, on the basis of a swift sort of *triage* that is accomplished *before* the information is passed on to those networks that complete the identification of the stimulus. (Yes, yes, I know. Only a person—a doctor or a nurse or such—can perform the behavior we call triage; I am speaking "metonymically." Get used to it.)

In conclusion, what I am telling my colleagues in the neurosciences is that there is no case to answer here. The authors claim that just about everybody in cognitive neuroscience is committing a rather simple conceptual howler. I say dismiss all the charges until the authors come through with some details worth considering. Do the authors offer anything else that might be of value to the neurosciences? They offer no positive theories or models or suggestions about how such theories or models might be constructed, of course, since that would be not the province of philosophy. Their "correct" accounts of commissurotomy and blindsight—for instance—consist in bland restatements of the presenting phenomena, not explanations at all. They are right so far as they go: that's how these remarkable phenomena appear. Now, how are we to explain them? Explanation has to stop somewhere, as Wittgenstein said, but not here. Bennett and Hacker quote with dismay some of the rudely dismissive remarks about philosophy by Glynn, Crick, Edelman, Zeki, and others (pp. 396–98). On the strength of this showing, one can see why the neuroscientists are so unimpressed.[23]

PUTTING CONSCIOUSNESS BACK
IN THE BRAIN

Reply to Bennett and Hacker, Philosophical
Foundations of Neuroscience

JOHN SEARLE

This is a long book, over 450 pages, and it covers a huge number of issues. It contains many objections to my views as well as an appendix specifically devoted to criticizing me. I will here confine my remarks to certain central issues in the book and to answering what I believe are the most important of Bennett and Hacker's criticisms. But I do not attempt to discuss all of the major issues raised by their book.

As most of my remarks will be critical, I want to begin by noting some important areas of agreement. The authors are right to point out that in perception we typically perceive actual objects in the world and not inner pictures or images of objects. They are also correct to point out that one's normal relation to one's own experiences is not epistemic. It is wrong to think of our relations to our perceptions on the model of "privileged access" or any other kind of epistemic "access," and the model of "introspection," whereby we *spect intro* our own minds, is hopelessly confused. I have made all these points myself in a number of writings and I am glad to see we are in agreement on these issues.[1]

I sent them the first draft of this paper and they kindly pointed out passages where they thought I had misunderstood or misstated their views, and this enabled me to correct certain misunderstandings.

Their basic conception of mental phenomena is, I believe, mistaken, and in this paper I will attempt to explain how. In order to state the difference between my views and theirs, I will begin with a brief summary of some of my views. I can then state their views clearly by contrast. For the sake of brevity I will confine my discussion to consciousness, though similar remarks could be made, *mutatis mutandis*, for intentionality.

Consciousness as a Biological Phenomenon

1. Consciousness, by definition, consists of states (I will use "states" to cover states, processes, events, etc.) that are *qualitative* and *subjective*. Pathologies apart, conscious states only occur as part of a *single unified conscious field*. Consciousness is qualitative in the sense that for any conscious state there is a certain qualitative character, a what-it-is-like or what-it-feels-like aspect. For example, the qualitative character of drinking beer is different from the qualitative character of listening to Beethoven's Ninth Symphony. These states are *subjective* in the ontological sense that they only exist as experienced by a human or animal subject. And they are *unified* in the sense that any conscious state, such as the present feel of the keyboard on my fingers, exists as part of one big conscious state, my present field of consciousness. Because of their subjective, qualitative character, these states are sometimes called "qualia." In general, I do not find this notion useful because it implies a distinction between conscious states that are qualitative and those that are not, and on my view there is no such distinction. "Consciousness" and "qualia" are simply

coextensive terms. However, because B and H deny the existence of qualia, I am going to use the term in this text to emphasize the points of disagreement. When I say conscious states exist, I mean qualia exist. When they say conscious states exist, they mean something else entirely, as we will see.

Conscious phenomena are concrete phenomena that go on in space-time. They are not abstract entities like numbers. I used to think that qualitativeness, subjectivity, and unity were three distinct features of consciousness, but, on reflection, it seems clear to me that each implies the next. These three are different aspects of the essence of consciousness: qualitative, unified subjectivity.

2. These states, qualia, are entirely caused by brain processes. We are not quite sure what the causal mechanisms are, but neuron firings at synapses seem to play an especially important functional role.

3. Conscious states exist in the brain. They are realized in the brain as higher level or system features. For example, conscious thoughts about our grandmothers are processes occurring in the brain, but, as far as we know, no single neuron can cause and realize thoughts about a grandmother. Consciousness is a feature of the brain at a level higher than that of individual neurons.[2]

There is of course a lot more to be said about consciousness and I have said some of it elsewhere. I take 1–3 to be more or less educated scientific common sense. But the interest for the present discussion is that, astounding as it may seem, *Bennett and Hacker deny all three.*

They claim to have shown that the notion of qualia, the notion of the qualitative character of conscious experiences, is "incoherent." They also say, "we part company with Searle ... when he claims that mental phenomena are caused by neurophysiological processes in the brain and are themselves features of the brain" (p. 446).

Just to see the magnitude of their denial and its consequences for philosophy and neuroscience, let us apply my three principles to an actual example. I now see a hand in front of my face. What are the component parts of the event of my seeing my hand? Well, first of all, there has to be a hand there, and it has to have a certain impact on my visual and neurobiological apparatus (I will spare you the details). In the normal case, such as the nonblindsight case, this impact will produce a conscious visual experience, in my sense of a qualitative, subjective event, a quale. I want to emphasize that the visual experience has all the features I just mentioned: it is qualitative, subjective, and exists as part of a unified field. It is caused by brain processes and it exists in the brain. So, we get three components in the visual scene: the perceiver, the object perceived, and the qualitative visual experience. Much of the very best work in neuroscience is an effort to explain how brain processes cause the visual experience and where and how it is realized in the brain. Astoundingly, Bennett and Hacker deny the existence of the visual experience in this sense, in the sense of quale. It is quite right to point out, as they do, that what I perceive is a hand and not a visual experience, but it is nothing short of bizarre to deny that there is a qualitative visual experience, in the sense of a visual quale, at all. What happens when I close my eyes, for example? The qualitative visual experience stops. That is why I stop seeing the hand, because I stop having the conscious visual experience. Notice that the presence of the hand is essential for my actually seeing the hand, but it is not essential for the existence of the visual experience because, in the case of hallucinations, I can have an indistinguishable experience without there being any hand there.

Bennett and Hacker are not the first authors to deny the existence of qualia, but their denial is not motivated by the mindless materialism of those who fear that if they grant the existence of irreducibly subjective mental phenomena they will find themselves in bed with Descartes. Well what does, what *could*, motivate a denial of the existence of conscious states, as I have defined them?

The Wittgensteinian Vision

The best way to understand their book is to see it as an application of Wittgenstein's philosophy of mind to contemporary neuroscience. Much of the originality of the book lies in the fact that this has not been done before. B and H's position is—as far as I know—unique in contemporary debates in the philosophy of mind.

What then is (their interpretation of) Wittgenstein's philosophy of mind, and how do they apply it to neuroscience? A key passage in Wittgenstein, which they quote, is this:

"Only of a living human being and what resembles (behaves like) a living human being can one say: it has sensations; it sees, is blind; hears, is deaf; is conscious or unconscious" (p. 71).

That is, it only *makes sense* to ascribe mental predicates to something that is, or behaves like, a living human being. And what is the role of the behavior in the ascription of these predicates? The behavior provides not just inductive grounds for the presence of mental phenomena, but logical criteria. We should see the behavioral manifestation as the logical criterion for the application of these concepts. There is a connection in meaning between the external behavior and the mental concepts, because only of a being that is capable of exhibiting a certain form of behavior can we say that it has a mental phenomenon. Two key sentences are these (p. 83): "*The criterial grounds of the ascription of a psychological predicate are partly constitutive of the meaning of that predicate*" and, same page, "*The brain does not satisfy the criteria for being a possible subject of psychological predicates.*"

Now, because Bennett and Hacker accept this Wittgensteinian conception, they think its immediate logical consequence is that consciousness cannot exist in brains, and mental activities, such as thinking and perceiving, cannot be performed by brains, because brains are incapable of exhibiting the appropriate behavior (p. 83). It is only of the whole person, or, in the case of animals, of the whole animal, that we can say it is in pain or is angry, because only the whole animal is capable of exhibiting the behavior that is partly constitutive of the conditions of application of the concept in question. Because of the criterial connection between mental states and behavior, we cannot make the traditional separation between the mental phenomenon and its external manifestation. The criterial connection explains why we can literally see that someone is angry or is in pain or is conscious or unconscious. Furthermore, Bennett and Hacker are now forced to deny the existence of qualia, for qualia, if they existed, would exist in brains, and that is inconsistent with the thesis that consciousness cannot exist in brains. For them, it is not just false that consciousness exists in brains, it is senseless. One might as well say that consciousness exists in prime numbers.

They also make a number of Wittgensteinian moves about how a child would learn the mental vocabulary and what the point is of having a mental vocabulary. They do not, as far as I remember, use the notion of a "language game," but it is implicit throughout the work. The language game that we play with the mental words requires publicly observable behavioral criteria for their application.

That is the vision, and I think that most of the substantive theses in the book really follow from that vision. Is this a valid basis on which to criticize contemporary work? Is it sufficient to refute the view that consciousness consists of unified, qualitative subjectivity, caused by brain processes and realized in brains? I think not, for the following reasons. Suppose they are right about the criterial basis and the necessity of publicly observable behavior for the language game to be played and the impossibility of a private language and all the rest of it. What follows? None of their spectacular conclusions follow. Once you grant, as you must, and as Wittgenstein did, that there is a distinction between the pain and the pain behavior, between the feelings of anger and the anger behavior, between the thought and the thought behavior, etc., then you can focus your neurobiological research attention on the pain, the feelings of anger, the visual experience, etc., and forget about the behavior. Just as the old-time behaviorists confused the behavioral evidence for mental states with the existence of the mental states themselves, so the Wittgensteinians make a more subtle, but still fundamentally similar, mistake when they confuse the criterial basis for the application of the mental concepts with the mental states themselves. That is, they confuse the behavioral criteria for the *ascription* of psychological predicates with the *facts ascribed by these* psychological predicates, and that is a very deep mistake.

Suppose Wittgenstein is right that we could not have the pain vocabulary unless there were common, publicly expressible forms of pain behavior. All the same, if I ask myself, "What fact about me makes it the case that I am in pain?" there is no fact about my behavior that makes it the case that I am in pain. The fact about me that makes it the case that I am in pain is the fact that I am having a certain sort of unpleasant sensation. And what goes for pain goes for anger, thinking, and all the rest of it. Even if the Wittgensteinian approach is 100 percent correct as a philosophical analysis of the operation of the *vocabulary*, all the same we can always carve off, in any individual case, the existence of the inner, qualitative, subjective feeling from its manifestation in external behavior. They point out that, though you can make the distinction in individual cases, it could not be the case that there never was any publicly observable manifestation of pain, for if so we would not be able to use the pain vocabulary. Let us suppose they are right about this. All the same, when we are investigating the ontology of pain—not the conditions for playing the language game, but the very ontology of the phenomenon itself—we can forget about the external behavior and just find out how the brain causes the internal sensations.

Notice that in the passage I quoted, Wittgenstein talks about what we can *say:*

"Only of a living human being and what resembles (behaves like) a living human being can one *say*: it has sensations; it sees, is blind; hears, is deaf; is conscious or unconscious" (p. 71; my emphasis).

But suppose we remove the word "say" from this passage and write it as follows:

"Only living human beings and things that behave like living human beings can actually *be* conscious." If we take this as a conceptual or logical claim, it is obviously false. Suppose we take it as,

"It is a conceptual or logical truth that only things that behave like human beings can actually be conscious."

But as a conceptual or logical truth, this just seems to be mistaken. For example, mollusks and crustaceans, such as oysters and crabs, do not behave at all "like human beings," but that fact by itself does not settle the question of whether they are conscious. Regardless of how their behavior differs from human behavior, oysters might still be conscious if they had the right sort of neurobiological processes in their nervous systems. Suppose we had a perfect science of the brain and we knew exactly how consciousness is produced in humans and the higher animals. If we then found that the consciousness–producing mechanism was present in oysters but not present in snails we would have very good reason, indeed, overwhelming reason, for supposing that oysters are conscious and snails are probably not conscious. The very existence of consciousness has nothing to do with behavior, even if, in humans, the existence of behavior is essential (criterial) for the operation of the language game. The question "Which of the lower animals are conscious?" cannot be settled by linguistic analysis.

Wittgenstein offers a general account of the functioning of the mental vocabulary. He points out that it is wrong to construe the operation of the language game on the model of external inductive evidence for the presence of inner private phenomena. "An inner process stands in need of an outward criterion," he reminds us. But, even if we accept that account of the vocabulary, there is nothing that prevents us from giving a neurobiological

account of how states of consciousness are caused by brain processes and realized in brain systems. Furthermore, the require-
ment that the system, the whole person, be capable of manifest-
ing behavior does not imply that there cannot be an element
of the system, the brain, that is the location of the conscious
processes. This is a separate point and I will explain it further in
the next section.

The fallacy, in short, is one of confusing the rules for using the
words with the ontology. Just as old-time behaviorism confused
the *evidence* for mental states with the *ontology* of the mental
states, so this Wittgensteinian criterial behaviorism construes the
grounds for making the attribution with the *fact that is attributed*. It is
a fallacy to say that the conditions for the successful operation
of the language game are conditions for the existence of the
phenomena in question. Suppose we had a perfect science of the
brain, so that we knew how the brain produces pain. Suppose we
made a machine that was capable of consciousness and indeed
capable of conscious pains. We might design the machine so that
it exhibited no pain behavior whatsoever. It would be up to us.
Indeed there are forms of actual illnesses where people have
pains without pain behavior. In some cases of the Guillain-Barré
syndrome the patient is completely conscious but totally para-
lyzed, totally unable to exhibit behavior corresponding to his or
her mental states. Bennett and Hacker point out that this could
not be the case for all pains, that no one in pain ever exhibited
pain behavior, because then we could not apply the words. Even
if that is right, it is a condition on the successful operation of the
language game, not a condition on the existence of pains.

Summary of the Argument So Far

I think that once this basic fallacy is removed, then the central
argument of the book collapses. I will discuss their detailed arguments in a moment, but now I want to summarize the arguments

so far. Wittgenstein claims that a condition of the possibility of a language for describing inner mental phenomena is publicly observable behavioral manifestations of those phenomena. The behavior is not just inductive evidence but is criterial for the application of the concepts. Suppose, for the sake of argument, that he is right. They think it follows that mental phenomena could not exist in brains, because brains cannot exhibit the criterial behavior. But this does not follow. All that could follow is that if we are to talk about mental states in brains then the brains must be part of a causal mechanism capable of producing behavior (I will discuss this point further in the next section). And normally they are. But even in cases where they are not, we need to distinguish the very existence of the mental phenomena from the possibility of talking about it. The fact about me that makes it the case that I have a pain when I do have a pain is the existence of a certain sort of sensation. Whether or not I manifest that sensation in behavior is irrelevant to its very existence.

The Main Argument of the Book:
The Mereological Fallacy

The single most important argument in the book, the one that they repeat over and over, is the exposure of what Bennett and Hacker call the "mereological fallacy," which they define as attributing to parts what only makes sense when attributed to the whole. The typical form that this fallacy takes, according to them, is that people say such things as that the *brain* thinks, perceives, hopes, wonders, decides, etc., when in fact the correct characterization is to say that the *whole person* thinks, perceives, hopes, wonders, decides, etc. The fallacy lies in attributing to the part, the brain, what only makes sense when attributed to the whole, the whole person. I hope it is obvious how this follows from the Wittgensteinian vision: because the conscious behavior cannot be exhibited by the part, i.e., the brain, and because the

conscious behavior is essential to the attribution of the consciousness, we cannot attribute pains to the brain.

I want to make a pedantic point, which will be important
later on. On their own account, this cannot be, strictly speaking,
a case of the mereological fallacy, the fallacy of attributing to the
part what only makes sense when attributed to the whole, because if it were, we could simply remove the fallacy by adding a
reference to the rest of the body. The relation of the brain to the
rest of the *body* is indeed part-whole. The brain is a part of my
body. They say only a *person* can be the subject of psychological
attributions, not just a brain. But the person is not related to the
brain as whole to part. That does not imply that the person is
something distinct from or "over and above" the body. Unfortunately, they never tell us what a person is, but I think it is crucial
to the whole account and indeed to the whole discussion. What
they call a mereological fallacy is, rather, a category mistake, in
Ryle's sense. On their account persons are in a different logical
category from brains, and for that reason, psychological attributions to persons do not make sense when attributed to brains. I
will also come back to this point later.

As they are aware, there are in fact (at least) three different
sorts of subpersonal attributions of psychological phenomena,
and arguments that they use against one do not necessarily apply to the others. First, there is brain as subject and agent (for
example, "The brain thinks"). Second, there is brain as location
of psychological processes (for example, "Thinking occurs in the
brain"). And, third, there are micro elements as agents (for example, "Individual neurons think"). Let us consider each of these
in order.

First, the brain as subject and agent. As I remarked above, it is
quite common both in the philosophical and in the neurobiological literature to describe cognition by using cognitive verbs
where the subject of the verb is "the brain." Thus, it is commonly said that the brain perceives, the brain thinks, the brain
decides, etc., and Bennett and Hacker find this unacceptable for

the reason I have stated: The brain cannot exhibit the appropriate behavior. In ordinary speech we have to say that the person decides. It is I who decided to vote for the Democratic candidate, not my brain.

The argument for this conclusion, as I said before, is that since the brain cannot exhibit behavior it cannot be the subject of psychological predicates. But once we see the weakness of that argument, can we think of any other reason for refusing to attribute psychological processes to the brain? I agree with them, that it is odd to say, for example, "My brain decided to vote for the democrats in the last election." Why? I will come back to this issue later.

Second, the brain as locus. A second form of attribution, which is really quite different from the first, is to state *where* the psychological processes and events actually occur. And here the claim is that they occur in the brain. B and H are aware of the distinction between treating the brain as the agent and subject of psychological processes and treating the brain as the locus of psychological processes; but they object to both. They think both that the brain cannot think, and that thinking cannot occur in the brain. But they would need a separate argument to show that the brain cannot be the locus of such processes, and I cannot find that argument. Strictly speaking, the Wittgensteinian argument, even if valid, would not hold against all of these attributions. Why not? The argument says that the agent of a psychological process must be a system capable of exhibiting the appropriate behavior. Thus, in our example of vision, a system capable of seeing must be able to exhibit the appropriate behavior. So we cannot say of a brain that it sees, we can only say it of the whole system, i.e., the person. But that does not preclude us from identifying the visual experience as a component of the seeing and locating the visual experience in the brain. All that the Wittgensteinian argument requires is that the brain be part of a causal mechanism of a total system capable of producing the behavior. And that condition can still be satisfied even when certain psychological processes are located in the brain.

To see this, consider an analogy. Suppose someone said: "We should not say that the stomach and the rest of the digestive tract digest food. Only the whole person can digest food." In a sense this is right. But notice that, for purposes of research into how it works, we can ask where and how the specific digestive processes occur. And the answer is that they occur in the stomach and the rest of the digestive tract. Now, similarly, someone might insist that it is me, the person, who consciously perceives and thinks, and not my brain. All the same, one can then ask where in the anatomy are the conscious processes occurring, and the obvious answer is that they are occurring in the brain. B and H are aware of the distinction but do not seem to be aware that the claim that the brain cannot be the locus of psychological processes would require a separate argument, and I do not find such an argument. They say, "The location of *the event of a person's thinking a certain thought* is the place where the person is when the thought occurs to him" (p. 180). No doubt that is true, but it does not imply that my thoughts do not also occur in my head. Certain thoughts are occurring to me right now. Whereabouts? In this room. Where exactly in this room? In my brain. Indeed, with the development of fMRI and other imaging techniques, we are getting closer to being able to say exactly where in my brain the thoughts occur.

The Location of Conscious States and the Causation of Those States by Brain Processes

The picture I have advanced is that conscious mental processes occur in the brain and are caused by lower-level neuronal processes. What is their picture? Once they deny that mental processes occur in the brain, I believe they are unable to give a coherent account of the location and causation of consciousness. They think that neural processes are a necessary condition of consciousness, but they do not, and, I believe, cannot, state the obvious point that, in appropriate circumstances, neurobiological processes are causally sufficient for consciousness. My present conscious states, qualia all, are caused by lower-level neuronal processes in the brain.

I think that some of the weakest arguments in the book are on this issue of the location and causation of conscious states. They say, "There is *no such thing* as a mental process (such as reciting the alphabet in one's imagination) going on in a part of an animal, no matter whether that part be the kidneys or the brain" (p. 112). "What goes on *in* the brain are neural processes, which need to occur in order for the person, whose brain it is, to be going *through* the relevant mental processes" (p. 112).

I believe this passage contains a deep mistake and I want to go through it step by step.

Suppose I recite the alphabet silently, as we would say "in my head." This is a real event in the real world. Like all real events it occurs in space-time. So, where did it happen? They say that I, the person, *went through* the mental process. No doubt that is right, but where exactly in real space-time did the conscious, datable, spatially located mental event of my silently reciting the alphabet occur? They cannot answer that question except to say such things as that it happened in New York or in this room. But that is not enough. I think it is obvious that a conscious event, a set of qualia, occurred inside my brain. And, to repeat, with advancing imaging techniques such as fMRI we are getting closer to saying exactly where it occurred.

This denial of the reality and spatial location of qualia prevents them from giving a coherent *causal* account of the relation of neural processes to mental events. They say that the neural processes are causally *necessary* ("need to occur") for a mental process to occur. But we need to know, in that context, what is causally *sufficient*, what made it happen that I "went through" a mental process? And whenever we talk about causes we need to say what exactly causes what exactly? On my account the conscious mental event is entirely caused by and realized in the brain. In that context those neuron firings are causally sufficient to produce those qualia, those conscious mental events. What is their account? They can't say that the neuron firings caused the qualia, the qualitative experience, because they have denied the existence of the qualia. They can't say the neuron firings caused the behavior, because there was no behavior. So what exactly is the nature of the mental process that I went through and what exactly are the causes that made it happen? The neurons fired . . . and then what? The neurons need to fire in order for me to "go through" the mental process. But what does going through the mental process consist in, if there are no qualia and no behavior? They have no answer to these questions, and, given their overall theory, I do not think they could have an answer.

On the Wittgensteinian assumption, if the brain cannot exhibit behavior then it cannot be the subject or agent of mental attributions. I believe that this is a mistake. But, whether or not it is a mistake, we still need to distinguish, in a way that Bennett and Hacker do not distinguish, between an argument that the brain cannot be the *subject* of psychological verbs and the argument that it cannot be the *locus* of psychological processes. They are aware of the distinction between brain as subject and brain as locus. They deny both that the brain can be a subject and that it can be the locus, but do not provide a separate argument against the claim that the brain is the locus of, for example, thought processes. Suppose we agreed that it sounds odd to say "My brain thinks . . . " All the same, when I *think* there can still be thought processes going on in my brain. The most the Wittgensteinian argument could establish is that we should not think of the brain as subject or agent. But it does not follow that it is not the *locus* of the corresponding processes. Their argument against *brain as subject* does not carry over into *brain as locus*.

The question, "Where do conscious though processes occur?" is no more philosophically puzzling than the question, "Where do digestive processes occur?" Cognitive processes are as much real biological processes as is digestion. And the answer is obvious. Digestion occurs in the stomach and the rest of the digestive tract; consciousness occurs in the brain and perhaps other parts of the central nervous system.

Metaphorical and Literal, Observer-Relative and Observer-Independent Attributions of Psychological States

Third, neurons as subjects and agents. A third form of attribution that Bennett and Hacker object to is when subcerebral parts of the brain are attributed psychological processes. Thus, for example, they quote Blakemore as saying that neurons perceive, neurons decide, neurons make inferences, etc. They think that this is also an instance of the mereological fallacy.

I take it that, properly understood, this is, or at least can be, a harmless metaphor. Indeed, in the scientific literature people make these sorts of attributions to the stomach. They say the stomach knows when certain sorts of chemicals are needed to digest certain sorts of carbohydrate inputs. As long as we keep clear the distinction between the literal observer-independent sense in which I infer or I receive information and the meta-phorical and observer-relative senses where we say my neurons make such and such inferences or my neurons perceive such and such phenomena, it seems to me that these metaphors are, or at least can be, harmless. It is easier to make the mistake of confus-ing the real observer-independent senses with the observer-rela-tive and metaphorical senses where the brain is concerned, than it is where other organs are concerned, for the obvious reason that intrinsic observer-independent psychological processes go on in the brain in a way that they do not go on in the stomach and the rest of the digestive tract. I am sure that Bennett and Hacker are right in thinking that at least some of the authors they criticize do not have a clear sense of the distinction be-tween the observer-independent attributions of these phenom-ena and the observer-relative and metaphorical attributions of these phenomena. The worst culprit, semantically speaking, is the notion of "information." An additional confusion is added by the engineers' use of "information" in the "information-theoretical" sense, which has nothing to do with information in the sense in which I have information that such and such is the case. We are told, for example, that the brain does information processing. Well, in one sense, that is obviously true. I take in information perceptually and I think about it, and then I derive new information by making inferences. The problem is that there are all sorts of subpersonal processes going on in the brain, in the lateral geniculate nucleus, for example, which can be described *as if* they were cases of thinking about information, but of course there is literally no information there. There are just neuron firings, which result in information of a conscious kind at the end of the process but themselves have no semantic content. Bennett and Hacker are clear about the distinction between the information theoretic sense of "information" and the intentionalistic sense. They refer to these as the "engineering" sense and the "semantic" sense, but I do not find in their book a clear statement of the distinction between the observer-independent sense and the observer-dependent sense of "information." I have observer-independent information about my phone number. The phone book has observer-dependent information about the same phone number. I have no objection to talking about information and information processing in the brain, provided that one is clear about these distinctions.

To summarize where we are so far: I have made three main objections to their argument. First the Wittgensteinian claim that behavior is criterial for the ascription of mental phenomena in a public language, even if true, does not refute the view that consciousness can exist in brains. Second, once we distinguish between brain as subject and brain as locus, B and H have no separate argument against brain as locus. In fact, as far as we know, it is true that the location of all our conscious processes is in the brain. Third, the attribution of psychological states to subpersonal entities such as neurons can be harmless as long as it is clear that is a metaphorical use. As long as one distinguishes the literal from the metaphorical, and the observer relative from the observer independent, there is nothing necessarily wrong with such attributions.

Their Arguments Against Qualia

In order to make the Wittgensteinian argument work, Bennett and Hacker have to have an independent argument against qualia. Why? Well, if qualia exist they would have to have a location, and the most obvious location is the brain. And that would be inconsistent with their overall theory. So I now turn to their arguments against qualia.

Consciousness consists, as I said earlier, of subjective, qualitative, unified mental processes that occur inside the physical space of the cranium in the actual human brain, presumably localized mostly in the thalamocortical system. I claim that our pains, tickles, and itches, for example, are *subjective,* in the sense that they can only exist insofar as they are experienced by an actual subject, and that they are *qualitative* and that they occur as *part of a unified conscious field.* Bennett and Hacker think they have counterarguments. First, using an argument that Ryle used to use, they say that the claim that each pain, tickle, or itch can only be experienced by a single subject is just a trivial grammatical claim with no ontological consequences. They say that, in exactly the same way, a smile has to always be someone's smile, or a sneeze has to be someone's sneeze, or, to use one of Ryle's examples, a catch made by a player in a game has to be someone's catch. In this sense the privacy of sneezes, smiles, and catches does not show anything ontologically significant. So they argue with pains, tickles, and itches. Yes, they have to be someone's pain, tickle, or itch, but that is just a trivial grammatical point, it does not give them any special status.

The answer to this, implicit in what I have already said, is that expressions referring to conscious states do not just have the *grammatical* feature of requiring a personal noun phrase for the identification of particular cases, but rather that the *subjectivity* of the phenomena themselves is tied to qualitativeness. And this is not just a grammatical point. There is a certain qualitative feel to a pain, a tickle, or an itch, and this is essential to the existence of pains, tickles, and itches. This qualitative feel is part of the ontological subjectivity in question, unlike the characteristics of catches, for example.

They offer independent arguments against the existence of qualia, but those arguments seem to me to be extremely weak. They say that the smell of a lilac and the smell of a rose will have the same qualitative character if they are equally pleasant or unpleasant (pp. 275–76). Astoundingly, they assume "quality" is a matter of degrees of pleasantness or unpleasantness. But this misses the point. The qualia of the smell of roses and the smell of lilacs is not constituted by the degrees to which they are pleasant or unpleasant. That is quite beside the point. The point is that the *character* of the experiences is different. This is what is meant by calling them qualia. Though, characteristically, qualia do have the features of pleasantness or unpleasantness, their defining essence is the *qualitative* feel of the experience. The answer to this point given by Bennett and Hacker also seems to me weak. Here is how it goes. They say that if you do not define qualia as a matter of pleasantness or unpleasantness, then you will have to individuate the experience by its object. It is a smell of a *lilac*, or it is a smell of a *rose*. And, they say, to identify the experience by its object is not to identify anything subjective about the experience, because roses and lilacs have an objective existence. Once again, this seems to me to miss the point. Of course, typically we identify the character of our perceptual experiences by their causes, that is to say, by the intentional object that causes us to have the characteristic experience. But one can have the experience and individuate the experience without the causes. If it turns out that my smell of the rose and my smell of the lilac were both hallucinations, that does not matter at all to the differences in the qualia. The qualitative aspects of the smell of a rose and the smell of a lilac remain the same whether there actually is an intentional object in either case. Often we discriminate smells without having any idea what caused them. It is, in short, not an objection to the existence of qualia with distinctive qualitative characters to say that typically when we describe qualia we describe them in terms of their objects, i.e., the phenomena that cause them. The notion of qualia can be defined in terms of the conscious experience by simply carving off the conscious experience from its intentional object. Qualia consists of the conscious experiences, however we choose to identify them. It is worth pointing out that chemists working in perfume factories try to synthesize chemicals that will duplicate the causal powers of such things as roses and lilacs. They are attempting to produce qualia that are type-identical qualia with those produced by the actual flowers.

Bennett and Hacker have a third argument that seems to me to also miss the point. They say that typically different people can feel the same pain or can have the same headache. If you and I both go to a party and drink too much wine, then the next day we will both have the same headache or, if we are both afflicted by the same disease that gives us a stomachache, we will have the same pain. Once again I think this misses the point that philosophers are trying to make about the "privacy" of pains. "Privacy" may be the wrong metaphor, but that is beside the point now. The point now is that what they are calling the same is the *type* and not the *token*. What we are interested in, when we talk about the privacy of pains, is not that different people cannot experience the same type of pain. Of course they can. But rather that the token pain that they or I experience *exists only as it is perceived* by a particular conscious subject.

The Location of Pains

I have said that all conscious states exist in the brain. But what about a pain in the foot? Surely it is in the foot and not in the brain. Bennett and Hacker object to my views on this question, and because my views are at first sight counterintuitive I want to make them clear. I believe that if we state all the facts clearly the questions concerning the location of pains will have obvious answers.

First, real space, that is, physical space. There is only one real, physical space, and everything in it is spatially related to everything else. Nowadays, after Einstein, we think of space and time as a single space-time continuum, and locations are specified relative to a coordinate system. For our purposes space is logically well behaved. Consider the transitivity of "in." If the chair is in the room and the room is in the house, then the chair is in the house. All events in the real world occur in physical space and time. Sometimes the boundaries of the events are ill defined—consider the Great Depression or the Protestant Reformation—for example, but, all the same, like all other events, they occur in space and time.

Now turn to experienced phenomenological body space. Suppose I receive an injury to my foot. This sets up a sequence of neuron firing that go up through my spine, through the tract of Lissauer, and into the pain centers of my brain, and, as a result, *I feel a pain in my foot.* There is no question that this is a correct description. So, for example, if asked by my doctor where the pain is, I point to my foot and not to my head. That is, to point to the location of the pain, I point to my anatomical limbs in real space. Our question now is, What exactly is the relation between real physical space and experienced phenomenological body space?

To answer that question we have to ask how the brain creates the phenomenological body space. The brain creates a body image, a phenomenological awareness of the parts of the body, their condition, and the relations between them. The brain creates within the body image an awareness of my foot and, consequently, when I feel the pain, an awareness of the pain as in my foot. We can summarize this by saying that the brain creates a phenomenologically real body space and a pain within the body space. There is no question about the phenomenological reality of these phenomena; the only question is, What exactly is the relation between phenomenological body space and the real physical space of my body?

Problems arise when we try to treat the phenomenological body space as if it were identical with the real physical space of the body. Notice that the transitivity of "in" does not work if we try to move from the phenomenological space to the physical space. I have a pain in my foot and my foot is in the room, but the pain is not in the room. Why not? The puzzle about the relation of phenomenological space to physical space becomes even more pressing when we consider phantom limb pains. The man feels a pain in his foot, but he has no foot. The pain is real, but where is it? Bennett and Hacker answer this question as follows. "So he actually has a pain where his foot would have been (i.e. in his phantom limb)" (p. 125). But a phantom limb is not a kind of limb, like an injured limb or a sunburned limb. Phantom limbs do not exist as objects in real space. So if we try to take their remark literally as about physical space, then it has the absurd result that the man "has a pain where his foot would have been" i.e., in the bed. The pain is right there under the sheet! Now why exactly is that absurd? Because in the physical space of beds and sheets there are no pains. Pains can only exist in phenomenological body spaces. And if we take the man's claim as about his phenomenological body image it is quite right. The man has a pain in his foot, even though he has no physical foot, he still has a phenomenological phantom foot in his body image.

But, and this the crucial point, the pain is a real event in the real world, so it must have a location in real space-time. It cannot be in his foot, because he does not have a foot. It cannot be where his foot would have been if he had had a foot, because there is nothing there between the sheets. It is of course in his phantom foot, but his phantom foot is not an object with a spatial location as part of the body like that of a real foot. I hope it is obvious that the spatial location in real world physical space of the phantom pain in the phantom foot is in the body image, which is in the brain. In real physical space both the pain in the real foot and the pain in the phantom foot are in the brain along with the rest of the body image.

Is a Person (an Embodied) Brain?

I now want to turn to the question I promised to discuss earlier, the apparent oddity of attributing psychological activities to the brain as well as to the whole person. I have disagreed with their arguments to show that brains cannot contain psychological processes, but I agree that it sounds odd to say, for example, "My brain decided to vote for the Democrats" Why does it sound odd? How is it that for some predicates we shift quite comfortably from discussing the person to discussing some feature such as the person's body? Consider the following four sentences.

1. I weigh 160 lbs.
2. I can visually discriminate blue from purple.
3. I have decided to vote for the Democrats.
4. I own property in the city of Berkeley.

In the case of 1 we have no hesitation in substituting for "I," "my body." That is, *I* weigh 160 lbs. if and only if *my body* weighs 160 lbs. The two claims seem to be equivalent. I also have no problems with the shift in 2. My brain, specifically my visual system, including the eyes, can discriminate blue from purple. But it seems to me that we are much more hesitant about making a similar shift in 3. If we say I have decided to vote for the Democrats, it seems more puzzling to say my brain or my embodied brain or the brain in my body has decided to vote for the Democrats. A corresponding shift for 4, I believe, is even more odd. If I say my embodied brain owns property in Berkeley, or my body owns property in

Berkeley, that sounds distinctly odd. Bennett and Hacker reject the shift for both 2 and 3. They would outlaw both "My visual system (in my brain) can visually discriminate blue from purple" and "My brain has decided to vote for the Democrats" because of the Wittgensteinian argument. I have already given reasons for rejecting the Wittgensteinian argument, but let us nonetheless grant that it does sound logically odd to say that my brain has decided to vote for the Democrats. Even if the Wittgensteinian argument is wrong, we have to address this oddity. Lots of neurobiologists and philosophers find it quite natural to attribute psychological activities to the brain.

How do we resolve this dispute? It seems to me that at the level we are discussing this now, there is a rather easy way to resolve the apparent dispute. Whenever you have an alternative wording for any statement *that S*, one initial way to approach the validity of the alternative wording is to ask yourself, what makes it the case *that p*, if *S* expresses the proposition *that p*? We have no trouble with 1 "I weigh 160 lbs." when we substitute "My body weighs 160 lbs." because we know what fact about me makes it the case that I weigh 160 lbs., namely, that that is what my body weighs. I have no objection to a similar shift for 2 because if we ask what fact about me makes it the case that I am able to distinguish blue from purple, the fact is that my visual system is able to distinguish blue from purple. But 3 and 4 are quite unlike 1 and 2 because they require not merely the existence and features of an embodied brain but rather that the embodied brain be socially situated and capable of social action. In the case of 3, unlike 4, we can carve off the social situation and identify a purely psychological component. Given my social and political situation, there are, in that context, certain activities going on in my brain that constitutes my having decided to vote for the Democrats. The same hesitation that makes us reluctant to attribute anything to the brain where the embodied person has to be socially situated also makes us reluctant to attribute any form of action or agency to the brain. Thus, though there are certain activities going on in my brain that constitute my having decided to vote for the Democrats, we are more reluctant to attribute this rational decision making to the brain than we are, for example, to attribute perceptual capacities to the brain. I have no trouble saying "my visual system can discriminate red from purple," but I am much more reluctant to say "my thalamocortical system has decided to vote for the Democrats." In the case of 4, "I own property in the city of Berkeley," there does not seem to be anything that we can carve off and attribute to the anatomy. It is only in virtue of my social situation and the relationships in which I stand that I can be a property owner. The property owner is indeed an embodied brain, but it is only under social and legal aspects that the embodied brain can be a property owner, and thus there is no possibility of making any anatomical or physiological attributions that are constitutive of the relevant portions of the facts.

These are interesting philosophical points, but I think the neurobiologists can simply sidestep them. Instead of worrying about to what extent they should treat rational agency as a feature of the neural anatomy, they should just go on with the second point, namely, it is just a fact that the psychological processes, which constitute conscious rational agency, are going on in the brain and can be investigated as such. For purposes of neurobiological research, the brain as causal mechanism and as anatomical location is enough.

I have discussed some of the philosophical points in other writings as part of the problem of the self.[3] Why do we need to postulate a self as something in addition to the sequence of our experiences and their anatomical realization? Not because there is some additional superanatomy or some additional superexperiences. There is just the embodied brain and the experiences that go on in that brain. Nonetheless, as I have argued, we do need to postulate a self, but it is a purely formal postulation. It is not an additional entity. It is a kind of principle of organization of the brain and its experiences.

The Nature of Philosophy

In the second appendix to the book, Bennett and Hacker criticize my views, both my account of problems in the philosophy of mind and my general approach to philosophy. They think I am mistaken in several respects in my conception of philosophy. In my experience, disputes about the nature of philosophy tend to be fruitless and usually just express preferences for different research projects. I was brought up philosophically in the Oxford of the 1950s where I was both a student and a don, and the prevailing orthodoxy was that philosophy was about language and about the use of words. If someone says that philosophy is entirely about language, that seems to me to express a preference. It is roughly equivalent to "I prefer philosophical work that is about language and I intend to pursue only philosophical work that is about language." What I discovered is that the techniques that I had used on language worked for other phenomena, specifically mental phenomena and social ontology. So the methods I use are a continuation of the methods of analytic philosophy, but extend far beyond the original domain of the philosophy of language and linguistic philosophy that I had been brought up on.

An essential disagreement between my approach and theirs is that they insist philosophy cannot be theoretical, that philosophy does not offer general theoretical accounts. We all agree that philosophy is, in some sense, essentially conceptual, but the question is, What is the nature of the conceptual analysis and what is the upshot? I can say from my own experience that a conceptual result is significant only as part of a general theory. So, if I look at my own intellectual history, I have advanced a general theory of speech acts and of meaning, a general theory of intentionality, a general theory of rationality, and a general theory of the nature of society, of social ontology. If somebody said, "Well, you cannot have theories in philosophy," my answer would be, "Just watch. Here are some general theories." A philosophical analysis of, for example, promising becomes much more powerful when it is incorporated into a general theory of language and speech acts. Philosophical analyses of action and perception become much deeper when they are embedded in a general theory of intentionality, and so on through the other cases.

In one respect they misstate my views on the relationship of philosophy and science. The claim I make is not that *all* philosophical problems can become scientific problems through careful conceptual analysis. On the contrary, I think that only a small number of philosophical problems admit of a solution in the natural sciences. The problem of life is one, and I hope that the so-called mind-body problem will become another. But most of the problems that worried the great Greek philosophers, such as, for example, the nature of the good life or the form of a just society or the best sort of social organization, I think are not the sort of things that, in any obvious way, are likely to be amenable to treatment by the natural sciences. So it is a misunderstanding for them to suppose that I think that all philosophical problems can eventually become scientific problems with scientific solutions. Such cases are exceptional.

Again, I have not found it possible to make a really sharp, precise distinction, as they claim to do, between empirical questions and conceptual questions, and consequently I do not make a sharp distinction between scientific and philosophical questions. Let me give one example to explain how my philosophical work can be helped by scientific discoveries. When I raise my arm, my conscious intention-in-action causes a physical movement of my body. But the movement also has a level of description where it is caused by a sequence of neuron firings and the secretion of acetylcholine at the axon end plates of the motor neurons. On the basis of these facts I can do a philosophical analysis to show that one and the same event must be both a qualitative, subjective, conscious event and also have a lot of chemical and electrical properties. But there the philosophical analysis ends. I need now to know how exactly it works in the plumbing. I need to know exactly how the brain causes and realizes the conscious intention-in-action in such a way that the intention, with its combination of phenomenological and electrochemical structures, can move physical objects. For that I am going to need the results of neurobiological research.

Bennett and Hacker have written a significant and in many ways useful book. They put a lot of work into it. I do not wish my objections to obscure its merits. However, I believe that the vision they present of neurobiology and the mind is profoundly mistaken and potentially harmful. Many of the crucial questions we need to ask in philosophy and neuroscience would be outlawed by their approach. For example, What are the NCCs (neuronal correlates of consciousness) and how exactly do they cause consciousness? How can my conscious intention-in-action move my body? Indeed, a huge number of central questions in neurobiological research would be rejected as meaningless or incoherent if their proposals were accepted. For example, the central question in vision, "How do neurobiological processes, beginning with the assault of the photons on the photoreceptor cells and continuing through the visual cortex into the prefrontal lobes, cause conscious visual experiences?" could not be investigated by anyone who accepted their conception. This is one of those cases, like strong AI, where a mistaken philosophical theory can have potentially disastrous scientific consequences, and that is why I consider it important to answer their claims.

REPLY TO THE REBUTTALS

THE CONCEPTUAL PRESUPPOSITIONS
OF COGNITIVE NEUROSCIENCE

A Reply to Critics

MAXWELL BENNETT AND PETER HACKER

Conceptual Elucidation

In *Philosophical Foundations of Neuroscience*[1] we aimed to contribute to neuroscientific research in the only way that philosophy can assist science—not by offering scientists empirical theories in place of their own, but by clarifying the conceptual structures they invoke. One of us has spent his life constructing empirical theories about neuronal functions. But those endeavors, which deal with the foundations of neuroscience, provide no part of its *conceptual* foundations. The systematic elucidations we gave of sensation, perception, knowledge, memory, thought, imagination, emotion, consciousness, and self-consciousness are not theories.[2] Their purpose is to clarify the psychological concepts that cognitive neuroscientists use in *their* empirical theories. The conceptual clarifications we gave demonstrate numerous incoherences in current neuroscientific theorizing. They show why these mistakes are committed and how to avoid them.

Cognitive neuroscience is an experimental investigation that aims to discover empirical truths concerning the neural foundations of human faculties and the neural processes that accompany their exercise. A precondition of truth is sense. If a form of words makes no sense, then it won't express a truth. If it does not express a truth, then it can't explain anything. Philosophical investigation into the conceptual foundations of neuroscience aims to disclose and clarify conceptual truths that are presupposed by, and are conditions of the sense of, cogent descriptions of cognitive neuroscientific discoveries and theories.[3] If conducted correctly, it will illuminate neuroscientific experiments and their description as well as the inferences that can be drawn from them. In *Philosophical Foundations of Neuroscience* we delineated the conceptual network formed by families of psychological concepts. These concepts are presupposed by cognitive neuroscientific research into the neural basis of human cognitive, cogitative, affective, and volitional powers. If the logical relations of implication, exclusion, compatibility, and presupposition that characterize the use of these concepts are not respected, invalid inferences are likely to be drawn, valid inferences are likely to be overlooked, and nonsensical combinations of words are likely to be treated as making sense.

Some philosophers, especially in the USA, have been much influenced by Quine's philosophy of logic and language, according to which there is no significant difference between empirical and conceptual truths.[4] So, from a theoretical point of view, a Quinean will hold that there is no essential difference between, for example, the sentence "Memory is knowledge retained" and the sentence "Memory is dependent upon the normal functioning of the hippocampus and neocortex." But this is wrong. The former expresses a conceptual truth, the latter a scientific discovery. According to Quine, the sentences of a theory face experience as a totality and are confirmed holistically. But it is mistaken to suppose that "vixens are female" is confirmed by the success of zoological theory or that "bachelors are unmarried" is confirmed by the sociology of marital habits. So, too, that red is darker than pink or that red is more like orange than like yellow is not verified by confirmation of the theory of color, but rather presupposed by it. It would be mistaken to suppose that the theorems of the differential calculus were confirmed holistically by the predictive success of Newtonian mechanics and might have been infirmed by its failure and rejection. They were confirmed by mathematical *proofs*. Nonempirical propositions, whether they are propositions of logic, mathematics, or straightforward conceptual truths, can be neither confirmed nor infirmed by empirical discoveries or theories.[5] Conceptual truths delineate the logical space within which facts are located. They determine what makes sense. Consequently facts can neither confirm nor conflict with them.[6]

A conceptual proposition ascribes internal properties or relations, an empirical proposition ascribes external ones. A conceptual truth is partly constitutive of the meanings of its constituent expressions, an empirical proposition is a description of how things stand. A conceptual truth is an implicit statement of a norm of description in the guise of a statement of fact. Precisely because such statements are partly constitutive of the meanings of their constituent expressions, failure to acknowledge a conceptual truth (e.g., that red is darker than pink) is a criterion for the lack of understanding of one or another of its constituent expressions.

This normative conception of conceptual truth is obviously not *especially* concerned with so-called analytic propositions under any of the various, familiar, *different* conceptions of analyticity, such as Kant's, Bolzano's, Frege's, and Carnap's. Indeed, the various analytic/synthetic distinctions are simply bypassed. Instead, we distinguish between the statement of a measure and the statement of a measurement. It would exhibit complete incomprehension to suppose that the distinction we are drawing between the conceptual and the empirical is an *epistemic* one.[7]

The distinction is *not* drawn by reference to *how we know* the respective truths. It is drawn by reference to the *role* of the proposition in question: whether it is normative (and constitutive) or descriptive. It should be emphasized that whether it is one or the other is a feature of the use of a sentence, not (or not necessarily) of a sentence type. A sentence used in one context to express a conceptual truth often may be used in another as a statement of fact—as is patent in Newtonian mechanics. In many contexts, it may be unclear *without further inquiry*, what role a sentence in use is meant to play. Indeed, it is typical, in science, for inductive evidence and constitutive evidence (logical criteria) to fluctuate. But this much is clear: to characterize a sentence as expressing a conceptual truth is to single out its distinctive *function* as a statement of a measure, rather than of a measurement. Hence the distinction, unlike the a priori/a posteriori distinction, is not epistemological, but logical or logico-grammatical.

Two Paradigms: Aristotle and Descartes

Philosophical reflection on human nature, on the body and soul, goes back to the dawn of philosophy. The polarities between which it fluctuates were set out by Plato and Aristotle. According to Plato, and the Platonic-Christian tradition of Augustine, the human being is not a unified substance but a combination of two distinct substances, a mortal body and an immortal soul. According to Aristotle, a human being *is* a unified substance, the soul (*psuchē*) being the form of the body. To describe that form is to describe the characteristic powers of human beings, in particular the distinctive powers of intellect and will that characterize the rational *psuchē*. Modern debate on this theme commences with the heir to the Platonic-Augustinian tradition, namely, the Cartesian conception of human beings as two one-sided things, a mind and a body. Their two-way causal interaction was invoked to explain human experience and behavior.

The greatest figures of the first two generations of twentieth-century neuroscientists, e.g., Sherrington, Eccles, and Penfield, were avowed Cartesian dualists. The third generation retained the basic Cartesian structure but transformed it into brain-body dualism: substance-dualism was abandoned, structural dualism retained. For neuroscientists now ascribe much the same array of mental predicates to the brain as Descartes ascribed to the mind and conceive of the relationship between thought and action, and experience and its objects, in much the same way as Descartes—essentially merely replacing the mind by the brain. The central theme of our book was to demonstrate the incoherence of brain/body dualism and to disclose its misguided crypto-Cartesian character. Our constructive aim was to show that an Aristotelian account, with due emphasis on first- and second-order active and passive abilities and their modes of behavioral manifestation, supplemented by Wittgensteinian insights that complement Aristotle's, is necessary to do justice to the structure of our conceptual scheme and to provide coherent descriptions of the great discoveries of post-Sherringtonian cognitive neuroscience.[8]

Aristotle's Principle and the Mereological Fallacy

In *Philosophical Foundations of Neuroscience* we identified a pervasive error that we called "the mereological fallacy in neuroscience."[9] Correcting this error is a leitmotiv (but *only* a leitmotiv) of our book. We called the mistake "mereological" because it involves ascribing to parts attributes that can intelligibly be ascribed only to the wholes of which they are parts. A form of this error was pointed out around 350 BC by Aristotle, who remarked that "to say that the soul [*psuchē*] is angry is as if one were to say that the soul weaves or builds. For it is surely better not to say that the soul pities, learns or thinks, but that a man does these with his soul" (DA 408^{b}12–15)—doing something with one's soul being like doing something with one's talents. It is mistaken to ascribe to the soul of an animal attributes that are properly ascribable only to the animal as a whole. We might call this "Aristotle's principle."

Our primary concern was with the neuroscientific cousin of this, namely, the error of ascribing to the *brain*—a part of an animal—attributes that can be ascribed literally only to the animal as a whole. We were not the first to have noted this—it was pointed out by Anthony Kenny in his brilliant paper "The Homunculus Fallacy" of 1971.[10] This error is more properly *mereological* than its Aristotelian ancestor, since the brain is literally a part of the sentient animal, whereas, contrary to the claims of Plato and Descartes, the soul or mind is not. In Aristotelian spirit we now observe that to say that the brain is angry is as if one were to say that the brain weaves or builds. For it is surely better to say not that the brain pities, learns, or thinks, but that a man does these.[11] Accordingly, we deny that it makes sense to say that the brain is conscious, feels sensations, perceives, thinks, knows, or wants anything—for these are attributes of animals, not of their brains.

We were a little surprised to find that Professor Dennett thinks that his distinction in *Content and Consciousness* of 1969 between personal and subpersonal levels of explanations is what *we* had in mind. He there wrote, correctly, that being in pain is not a property of the brain. But his reason was that pains are "mental phenomena" that are "non-mechanical," whereas cerebral processes are "essentially mechanical."[12] The contrast *we* drew between properties of wholes and properties of parts is not between what is nonmechanical and what is mechanical. It is the bracket clock as a whole that keeps time, not its fusée or great wheel—although the process of keeping time is wholly mechanical. It is the aeroplane that flies, not its engines—although the process of flying is wholly mechanical. Moreover, verbs of sensation such as "hurts," "itches," "tickles" *do* apply to the parts of an animal, whose leg may hurt, whose head may itch, and whose flanks may tickle (PFN 73). These attributes are, as Professor Dennett puts it, "non-mechanical"; nevertheless they *are* ascribable to parts of an animal. So the mereological point we made is quite different from Professor Dennett's distinction between personal and subpersonal levels of explanation, and, applied to animals, is quite different from his distinction between what is "mechanical" and what is not.[13]

Is the Mereological Fallacy Really Mereological?

Professor Searle objected that what we characterize as a paradigm of a mereological fallacy, i.e., the ascription of psychological attributes to the brain, is no such thing, for the brain is not a part of a person, but rather a part of a person's body (p. 107). This, we think, is a red herring. The dictum of Wittgenstein that we quoted was "Only of a *human being* and of what resembles (behaves like) a living human being can one say: it has sensations; it sees, is blind; hears, is deaf; is conscious or unconscious"[14] (our emphases). The brain *is* a part of the human being.

Professor Searle suggests that if ascribing psychological attributes to the brain really were a mereological error, then it would vanish if one ascribed them to what he calls "the rest of the system" to which the brain belongs. He thinks that the "rest of the system" is the body that a human being *has*. He observes that we do not ascribe psychological predicates to the body one has. With the striking exception of verbs of sensation (e.g., "My body aches all over"), the latter point is correct. We do not say "My body perceives, thinks, or knows." However, "the system" to which the human brain can be said to belong is *the human being*. The human brain is a part of the human being, just as the canine brain is a part of a dog. My brain—the brain I have—is as much a part of me—of the living human being that I am—as my legs and arms are parts of me. But it is true that my brain can also be said to be a part of my body.

How is this to be explained? Our talk of our mind is largely *nonagential, idiomatic* talk of our rational powers of intellect and will, and of their exercise. Our talk of our body is talk of our corporeal properties. To speak of *my body* is to speak of corporeal features *of the human being that I am*—features pertaining to appearance (an attractive or ungainly body), to the superficies of the human being (his body was covered with mosquito bites, was lacerated all over, was painted blue), to aspects of health and fitness (a diseased or healthy body), and, very strikingly, to sensation (my body may ache all over, just as my leg may hurt and my back may itch).[15] But knowing, perceiving, thinking, imagining, etc. are not corporeal features of human beings and are not ascribable to the body a human being has, any more than they are ascribable to the brain that a human being has. Human beings are not *their bodies*. Nevertheless they *are* bodies, in the quite different sense of being a particular kind of sentient spatiotemporal continuant—*homo sapiens*—and the brain is a part of the living human being, as are the limbs.[16] It is not, however, a conscious, thinking, perceiving part—and nor is any other part of a human being. For these are attributes of the human being as a whole.

Nevertheless, Professor Searle has noted an interesting feature of our corporeal idiom. Human beings are persons—that is, they are intelligent, language-using animals—are self-conscious, possess knowledge of good and evil, are responsible for their deeds, and are bearers of rights and duties. To be a person is, roughly speaking, to possess such abilities as qualify one for the status of a moral agent. We would probably not say that the brain is part of the person but rather that it is part of the person's body, whereas we would not hesitate to say that Jack's brain is a part of Jack, part of *this* human being, just as his legs and arms are parts of Jack. Why? Perhaps because "person" is, as Locke stressed "a forensic term," but not a substance-name. So, if we use the term "person" in contexts such as this, we indicate thereby that we are concerned primarily with human beings qua possessors of those characteristics that render them persons, in relative disregard of corporeal characteristics. Perhaps the following analogy will help: London is a part of the UK; the UK belongs to the European Union, but London does not. That does not prevent London from being part of the UK. So too Jack's being a person does not prevent his brain being part of him.

The Rationale of the Principle

Why should one accept Aristotle's principle and its neuroscientific heir? Why should we discourage neuroscientists from ascribing *consciousness, knowledge, perception,* etc. to the brain?

Consciousness. It is animals that are conscious or unconscious, and that may become conscious of something that catches their attention. It is the student, not his brain, who awakes and becomes conscious of what the lecturer is talking about, and it is the lecturer, not his brain, who is conscious of his students' boredom as they surreptitiously yawn. The brain is not an organ of consciousness. One sees with one's eyes and hears with one's ears, but one is not conscious with one's brain any more than one walks with one's brain.

An animal may be conscious without showing it. That is the *only* sense in which one can say, with Professor Searle, that "the very existence of consciousness has nothing to do with behavior" (p. 104). But, *the concept* of consciousness is bound up with the behavioral grounds for ascribing consciousness to the animal. An animal does not have to exhibit such behavior in order for it *to be* conscious. But only an animal to which such behavior *can intelligibly be ascribed* can also be said, *either truly or falsely*, to be conscious. It makes no sense to ascribe consciousness or thought to a chair or an oyster, because there is no such thing as a chair or oyster falling asleep and later waking up, or losing consciousness and then regaining it again; and there is no such thing as a chair or oyster behaving thoughtfully or thoughtlessly.[17] The "ontological question" (as Professor Searle puts it)—the question of

truth (as we should prefer to put it)—presupposes the antecedent determination of the question of sense. Agreement on the
behavioral grounds for ascription of consciousness, i.e., on what
counts as a manifestation of consciousness, is a precondition for
scientific investigation into the neural conditions for being conscious. Otherwise one could not even identify what one wants
to investigate. To distinguish the question of sense from the question of truth is not to confuse "the rules for using words with the
ontology," as Professor Searle suggested (p. 105)—on the contrary,
it is to distinguish them.[18]

Professor Searle insists that consciousness is a property of the
brain. Sherrington, Eccles, and Penfield, being Cartesians, wrong-
ly thought it to be a property of the mind. What recent neuroscientific experiment can Professor Searle cite to show that it is
actually a property of the brain? After all, the only thing neuroscientists *could* discover is that certain neural states are inductively
well correlated with, and causal conditions of, an *animal's* being
conscious. But *that* discovery cannot show that it is *the brain* that
is conscious. Is Professor Searle's claim then a conceptual insight?
No—for that is not the way the concept of *being conscious* is deployed. It is human beings (and other animals), not their brains (or
their minds), that fall asleep and later awaken, that are knocked
unconscious and later regain consciousness. So is it a linguistic
recommendation: namely, that when a human being's brain is in
a state that is inductively well-correlated with the human being's
being conscious, we should describe his brain as being conscious
too? This is a convention we could adopt. We could introduce this
derivative use of "to be conscious." It is necessarily parasitic on the
primary use that applies to the human being as a whole. It is, however, difficult to see anything that recommends it. It is certainly
not needed for the sake of clarity of description, and it adds nothing but an empty form to existing neuroscientific explanation.

Knowledge. Knowledge comprises abilities of various kinds.
The identity of an ability is determined by what it is an ability
to do. The simplest grounds for ascribing an ability to an animal

is that it engages in corporeal activities that manifest its abilities. The more complex the ability, the more diverse and diffuse the grounds. If an animal knows something, it can act and respond to its environment in ways that it cannot if it is ignorant; if it does so, it manifests its knowledge. The brain can be said to be the *vehicle* of these abilities, but what this means is that in the absence of the appropriate neural structures the animal would not be able to do what it can do. The neural structures in the brain are distinct from the abilities *the animal* has, and the operations of these structures are distinct from the exercise of the abilities *by the animal*. In short, the knower is also the doer, and his knowing is exhibited in what he does.

We criticized J. Z. Young for holding, as many neuroscientists do, that the brain contains knowledge and information "just as knowledge and information can be recorded in books or computers."[19] Professor Dennett avers that we did nothing to establish that there is no concept of knowledge or information such that it cannot be said to be encoded in both books and brains (p. 91). In fact we did discuss this (PFN 152f.). But we shall explain again.

A code is a system of encrypting and information transmission conventions parasitic on language. A code is not a language. It has neither a grammar nor a lexicon (cf. Morse code). Knowledge is not *encoded* in books, unless they are written in code. One can encode a message only if there is a code in which to do so. There is a code only if encoders and intended decoders agree on encoding conventions. In this sense there isn't, and couldn't be, a neural code. In the sense in which a book contains information, the brain contains none. In the sense in which a human being possesses information, the brain possesses none. That information can be derived from features of the brain (as dendrochronological information can be derived from a tree trunk) does not show that information is encoded in the brain (any more than it is in the tree trunk).

So, in the ordinary sense of "knowledge," there can be no knowledge recorded, contained in or possessed by the brain.

Professor Dennett then changes tack and recommends that we attend to the cognitive scientific literature on *extensions* of the term "knowledge" that might allow knowledge, in an extended sense, to be ascribed to the brain. And he recommends to our attention Chomsky's attempt to explain an extended concept of knowledge, namely, "cognizing," according to which human beings, and even neonates, cognize the principles of universal grammar.[20] According to Chomsky, someone who cognizes cannot tell one what he cognizes, cannot display the object of his cognizing, does not recognize what he cognizes when told, never forgets what he cognizes (but never remembers it either), has never learnt it, and could not teach it. Apart from that, cognizing is just like knowing! Does *this* commend itself as a model for an intelligible extension of a term?

Perception: The perceptual faculties are powers to acquire knowledge by the use of one's sense organs. An animal uses its eyes in glancing at, watching, peering at, and looking at things. It is thus able to discriminate things that are colored, that have distinctive shapes and movements. It exhibits its visual acumen in what it does in response to what it sees. It would not have these perceptual powers or be able to exercise them but for the proper functioning of appropriate parts of its brain. However, it is not the cerebral cortex that sees, but the animal. It is not the brain that moves closer to see better, looks through the bushes and under the hedges. It is not the brain that leaps to avoid a predator seen or charges the prey it sees—it is the perceiving animal. In short, the perceiver is also the actor.

In *Consciousness Explained* Professor Dennett ascribed psychological attributes to the brain. He asserted that it is conscious, gathers information, makes simplifying assumptions, makes use of supporting information, and arrives at conclusions.[21]

This is to commit the very fallacy that both Aristotle and Wittgenstein warned against—the mereological fallacy, as we called it. In his APA paper, Professor Dennett concedes that it would be a fallacy to attribute *fully fledged* psychological predicates to parts of the brain (p. 87). Nevertheless, he holds, it is theoretically fruitful, and consistent with accepting the erroneous character of attributing predicates of wholes to their parts, to extend the psychological vocabulary, *duly attenuated*, from human beings and other animals to a. computers and b. parts of the brain. Indeed, he apparently holds that there is no difference of moment between these two extensions. But there is a difference. Attributing psychological properties (attenuated or otherwise) to computers is mistaken, but does not involve a mereological fallacy. Attributing such psychological properties to the brain or its parts is mistaken and does involve a mereological fallacy. Taking the brain to be a computer and ascribing such psychological properties to it or its parts is doubly mistaken. We shall explain.

It is true that we do, in casual parlance, say that computers remember, that they search their memory, that they calculate, and sometimes, when they take a long time, we jocularly say that they are thinking things over. But this is merely a *façon de parler*. It is not a literal application of the terms "remember," "calculate," and "think." Computers are devices designed to fulfill certain functions for us. We can store information in a computer, as we can in a filing cabinet. But filing cabinets cannot remember anything, and neither can computers. We use computers to produce the results of a calculation—just as we used to use a slide-rule or cylindrical mechanical calculator. Those results are produced without anyone or anything literally calculating—as is evident in the case of a slide rule or mechanical calculator. In order literally to calculate, one must have a grasp of a wide range of concepts, follow a multitude of rules that one must know, and understand a variety of operations. Computers do not and cannot.

Professor Dennett suggests that "it is an empirical fact . . . that *parts* of our brains engage in processes that are *strikingly like* guessing, deciding, believing, jumping to conclusions, etc. and it is *enough* like these personal level behaviors to warrant stretching ordinary usage to cover it" (p 86). He agrees that it would be mistaken to "attribute *fully fledged* belief," decision, desire, or pain to the brain. Rather, "Just as a young child can *sort of* believe that her daddy is a doctor . . . so . . . some part of a person's brain can *sort of* believe that there is an open door a few feet ahead" (p. 87).

This is part of what Professor Dennett characterizes as "the intentional stance"—a research methodology that supposedly helps neuroscientists to explain the neural foundations of human powers. He claims that adoption of the intentional stance has accomplished "excellent scientific work . . . generating hypotheses to test, articulating theories, analysing distressingly complex phenomena into their more comprehensible parts" (p. 87). It seems committed to the idea that some parts of the brain "sort of believe," that others *sort of decide*, and yet others *sort of oversee* these activities. All this, presumably, is supposed to *sort of explain* what neuroscientists want to explain. But if the explananda are uniformly sorts of believings, pseudo-expectings, proto-wantings, and demi-decidings (as Dennett suggests [p. 88]), they at best only *sort of make sense* and presumably are only *sort of true*. And how one can make valid inferences from such premises is more than just sort of obscure. How precisely such premises are supposed to *explain* the phenomena is equally obscure. For the logic of such putative explanations is altogether unclear. Does sort of believing, pseudo-believing, proto-believing, or demi-believing something furnish a part of the brain with a reason for acting? Or only a sort of reason?—for a sort of action? When asked whether parts of the brain are, as Dennett puts it, "real intentional systems," his reply is "Don't ask" (p. 88).[22]

Cognitive neuroscientists ask *real* questions—they ask *how* the prefrontal cortices are involved in human thinking, *why* reentrant pathways exist, *what* precisely are the roles of the hippocampus and neocortex in a human being's remembering. Being told that the hippocampus sort of remembers for a short while and that the neocortex has a better sort of long-term memory provides no explanation whatsoever. No well-confirmed empirical theory in neuroscience has emerged from Dennett's explanations, for ascribing "sort of psychological properties" to parts of the brain does not *explain* anything. We shall revert to this when we discuss Sperry and Gazzaniga's account of commissurotomy. Not only does it not explain, it generates further incoherence.[23]

We agree with Professor Dennett that many of a child's beliefs are beliefs in an attenuated sense. A little girl's grasp of the concept of a doctor may be defective, but she will rightly say "Daddy is a doctor" and reply to the question "Where is the doctor?" by saying "In there" (pointing to Daddy's office). So she can be said to believe, in an attenuated sense, that her father is a doctor. She satisfies, in her verbal and deictic behavior, *some* of the normal criteria for believing that her father is a doctor (but also satisfies some of the criteria for lacking this belief). But there is no such thing as a part of a brain asserting things, as the child does, answering questions, as the child does, or pointing at things, as the child does. So in the sense in which, in her verbal and deictic behavior, the child can manifest *rudimentary* belief, a part of a brain can no more do so than the whole brain can manifest fully fledged belief. Or can Professor Dennett suggest an *experimentum crucis* that will demonstrate that her prefrontal cortex sort of believes that the cat is under the sofa?

The child can also exhibit rudimentary belief in her nonverbal behavior. If she sees the cat run under the sofa and toddles over to look for it, then she can be said to think the cat is under the sofa. But brains and their parts cannot *behave*, cannot toddle over to the sofa, cannot look under it, and cannot look nonplussed when there is no cat there. Brain parts can neither voluntarily act nor take action. Unlike the child, brain parts cannot satisfy *any* of the criteria for believing something, even in a rudimentary sense. Brains (and their parts) can only "sort of believe" in the sense in which they are "sort of oceans" (since there are brain waves) and are "sort of weather systems" (since there are brainstorms). The similarity between a brain and an ocean is at least as great as the similarity of brain processes to human beings' believings, decidings, or guessings. After all, both brains and oceans are gray, have wrinkles on their surface, and have currents running through them.

The Location of Psychological Attributes

The question of whether the brain is a possible subject of psychological attributes is distinct from the question whether the brain is the locus of those psychological attributes to which a corporeal location can intelligibly be assigned (PFN 122f., 179f.). Our reasons for denying that the brain can be the subject of psychological attributes do not show that the brain is not the locus of such attributes to which it makes sense to assign a corporeal location. Nor were they meant to. Our view is that sensations such as pains and itches *can* be assigned a location. The location of a pain is where the sufferer points, in the limb that he assuages, in the part of his body he describes as hurting—for it is these forms of pain behavior that provide criteria for the location of pain. By contrast, thinking, believing, deciding, and wanting, for example, cannot be assigned a *somatic* location. The answer to the questions "Where did you think of that?" "Where did he acquire that strange belief?" "Where did she decide to get married?" is *never* "In the prefrontal lobes, of course." The criteria for where a human being thought of something, acquired a belief, made a decision, got angry or was astonished involve behavior, to be sure, but not somatic-location-indicative behavior. The location of a human being's thinking, recollecting, seeing, deciding, getting angry, or being astonished is *where the human being is when he thinks*, etc.[24] Which part of his brain is involved in his doing so is a further, important question about which neuroscientists are gradually learning more. But they are not learning where thinking, recollecting, or deciding occur—they are discovering which parts of the cortex are causally implicated in a human being's thinking, recollecting, deciding.

Of course, thinking about something, deciding to do something, seeing something, are, as Professor Searle rightly said (p. 110), real events—they really happen somewhere, some when, in the world. I thought up that argument in the library and decided how to phrase it in my study; I saw Jack when I was in the street and I listened to Jill's recital in the concert hall. Professor Searle suggested that the question "Where do mental events occur?" is no more philosophically puzzling than the question "Where do digestive processes occur?" So, he argued, digestive processes occur in the stomach, and consciousness occurs in the brain. This is mistaken. Being conscious, as opposed to unconscious, being conscious of something, as opposed to not noticing it or not attending to it, do not occur *in* the brain at all. Of course, they occur *because of certain events in the brain*, without which a human being would not have regained consciousness or had his attention caught. "Where did you become conscious of the sound of the clock?" is to be answered by specifying where I was when it caught my attention, just as "Where did you regain consciousness?" is to be answered by specifying where I was when I came round.

Both digesting and thinking are predicated of animals. But it does not follow that there are no logical differences between them. The stomach can be said to be digesting food, but the brain cannot be said to be thinking. The stomach is the digestive organ, but the brain is no more an organ of thought than it is an organ of locomotion.[25] If one opens the stomach, one can see the digestion of the food going on there. But if one wants to see thinking going on, one should look at the *Le Penseur* (or the surgeon operating or the chess player playing or the debater debating), not at his brain. All his brain can show is what goes on there *while he is thinking*; all fMRI scanners can show is which parts of his brain are metabolizing more oxygen than others when the patient in the scanner is thinking.[26] (We ascribe length, strength, and having cracks to steel girders. But it does not follow that *length* and *strength* have the same logical character; and one can ask where the crack is, but not where the strength is.)

So, sensations, such as pains, *are* located in our bodies. But Professor Searle holds that they are all *in the brain*. It is, he admits, counterintuitive—after all we complain of stomachache, of gout in our foot, or arthritis in our knees. Nevertheless, he claims, the brain creates a body image, and the pain that we describe as being in the foot, and which we assuage by rubbing the foot, is an awareness-of-the-pain-as-in-my-foot, which is in the body image that is in one's brain. It is interesting that Descartes took a very similar view, remarking that "pain in the hand is felt by the soul not because it is present in the hand but because it is present in the brain."[27] The advantage of his account, Professor Searle suggests, is that it means that we can describe the phenomenon of phantom pain without the absurdity of suggesting that the pain is in physical space, in the bed or underneath the sheet. But that absurdity, he holds, is what we are committed to by claiming that pains are in the body. We agree on the absurdity, but deny that we are committed to it.

There are many locative uses of "in," some spatial, some non-spatial ("in the story," "in October," "in committee"). Among spatial uses, there are many different kinds, depending on what is in what (PFN 123f.). We agree with Professor Searle that if there is a coin in my jacket pocket, and if my jacket is in the dresser, then there is coin in the dresser. But not all spatial locative uses of "in" are thus transitive. If there is a hole in my jacket and the jacket is in the wardrobe, it does not follow that there is a hole in the wardrobe. In the case of the jacket and the coin, we are concerned with spatial relations between two independent objects, but not in the case of the jacket and the hole. Similarly, if there is a crease in my shirt, and my shirt is in the suitcase, it does not follow that there is a crease in the suitcase. The coin may be taken out of the jacket pocket, and the shirt may be taken out of the suitcase, but the hole cannot be taken out of the pocket—it has to be sewn up, as the crease has to be ironed out, not taken out.

The use of "in" with respect to the location of sensations is not like the coin, but more like the hole (though still different). A pain is not a substance. If I have a pain in my foot, I do not stand in any *relation* to a pain—rather, my foot hurts *there*, and I can point to the place that hurts, which we call "the location of the pain." In the case of the phantom limb, it feels to the sufferer just as if he still has the limb that has been amputated, and he avows a pain in the illusory limb. It seems to him just as if his leg were hurting, although he has no leg. We agree with Professor Searle that it is not the bed that hurts nor is the pain the amputee feels under the sheet. That he feels the pain where his leg would have been, and that his leg would have been under the sheet, do not imply that there is a pain under the sheet, any more than his having a pain in his unamputated leg and his leg being in his boot implies that he has a pain in his boot. Indeed, we agree with Professor Searle about the phenomena, and disagree only over its description. That the amputee's pain is real but its felt location is illusory (his leg does not hurt, as he has no leg) does not show that when a person who has *not* suffered an amputation feels a pain in his leg, its felt location is illusory too. It really is his leg that hurts! We do not think that there are body images in the brain and wonder what evidence there is for their existence—after all, one cannot find body images if one opens up the brain of a human being. What Professor Searle is apparently referring to is that physiological methods, beginning with those of Sherrington, have been used to establish that neurons in the somatosensory cortex can be excited in a topographical one-to-one relation with points stimulated on the surface of the body and with the spatial layout of the muscles of the limbs and trunk. But it is altogether unclear what Professor Searle means by "having a pain in a phenomenological phantom foot in a body image in the brain" (p. 118f.). One *can* have pains in one's *head*—they are commonly known as headaches. But one cannot have a backache or a stomachache in one's brain, or any other pain. And that is no coincidence, since there are no fiber endings there save in the dura.

Finally, Professor Searle claims that when philosophers say that two people both have the same pain, what they mean is that they have the same type pain, but different token pains (p. 116). "The token pain that they or I experience *exists only as it is perceived* by a particular conscious subject" (p. 116). This is mistaken. First of all, pains are not *perceived* by their sufferers. To *have* a pain is not to *perceive* a pain. "I have a pain in my leg" no more describes a relation between me and an object called "a pain" than does the equivalent sentence "My leg hurts." Second, Peirce's type/token distinction was applied to inscriptions and is dependent on or-thographic conventions. It no more applies to pains than it does to colors. If two armchairs are both maroon, then there are two chairs of the very same color, and not two token colors of the same type. For how is one to individuate the different tokens? Obviously not by location—since that merely distinguishes the two colored chairs, not their color. All one can say is that the first alleged token belongs to the first chair and the second to the second chair. But this is to individuate a property by refer-ence to the pseudo-property of belonging to the substance that has it—as if properties were substances that are distinguished by means of Leibniz's law and as if being the property of a given substance were a property that distinguishes, for example, the color of *this* chair from the color of *that* one. And that is absurd. The two chairs are both of the very same color. Similarly, if two people have a splitting headache in their left temples, then they both have the very same pain. A's pain is not distinguishable from B's pain by virtue of the fact that it belongs to A, any more than the maroon color of the first chair is distinguished from the maroon color of the second chair by virtue of the fact that it belongs to the first. The distinction between qualitative and nu-merical identity does not apply to colors or to pains, and neither does the Peircean distinction between types and tokens.

Linguistic Anthropology, Autoanthropology, Metaphor, and Extending Usage

Professor Dennett suggests that to examine the use of words in-volves either a form of anthropology or a form of "autoanthro-pology." For one has to discover the uses of words by doing ap-propriate social surveys, asking people to consult their intuitions on correct word usage. Alternatively, one has to consult one's own intuitions; but then it might turn out that one's intuitions diverge from those of others. He avers that we did not consult the community of neuroscientists to discover their intuitions about their neuroscientific patois of psychological predicates (p. 86 and fn. 15, p. 204), but only our own intuitions.

This is a misconception. A competent speaker of the language no more has to consult his intuitions (hunches, guesses) than a competent mathematician has to consult his intuitions concerning the multiplication tables or a competent chess player has to consult his intuitions about the movements of chess pieces. It is an empirical fact, to be established by anthropologists, historical linguists, etc., that a given vocable or inscription is or was used in a certain way in a given linguistic community. It is not an empirical fact that a word, *meaning what it does*, has the conceptual connections, compatibilities, and incompatibilities that it does. It is an empirical fact that the vocable "black" is used by English speakers to mean what it does, but given that it means what it does, namely, *this* ☞ ■ color, it is not an empirical fact that the propositions "Black is darker than white," "Black is more like gray than like white," "Nothing can be both black all over and white all over simultaneously" are true. These are conceptual truths specifying a part of the conceptual network of which *black* is a node. Failure to acknowledge these truths betokens a failure fully to have grasped the meaning of the word. A competent speaker is one who has mastered the usage of the common expressions of the language. It is not an intuition of his that black is *that* ☞ ■ color, that a vixen is a female fox, or that to perambulate is to walk. It is not a hunch of his that a man is an adult male human being. And it is no guess of his that if it is ten o'clock it is later than nine o'clock or that if something is black all over it is not also white all over.

Although competent speakers of a language agree in the language they use, deviations from common usage are not, *as such*, philosophically pernicious. Such deviations may betoken no more than a fragment of a personal idiolect or a special sociolect, a novel extension of a term or the appropriation of an existing term for a new technical use. That is why we wrote that if a competent speaker uses expressions contrary to usage, then it may well be that

> his words must not be taken to have their ordinary meaning. The problematic expressions were perhaps used in a special sense, and are really homonyms; or they were analogical *extensions* of the customary use, as is indeed common in science; or they were used in a metaphorical or figurative sense. If these escape routes are available, then the accusation that neuroscientists fall victim to the mereological fallacy is unwarranted.   (PFN 74)

But that these escape routes are available is not a matter that can be taken for granted. Nor is the cogency of this application of the term a matter on which the speaker in question is the final authority. For even if he is introducing a new use, or employing his words figuratively, whether he does so coherently has to be seen. And whether he does so consistently or rather moves unawares between a new use and an old one, drawing inferences from the former that are licensed only by the latter, has to be investigated. That is why we wrote

> The final authority on the matter is *his own reasoning*. We must look at the consequences he draws from his own words—and it is his inferences that will show whether he was using the predicate in a new sense or misusing it. If he is to be condemned, it must be out of his own mouth.
>
> (PFN 74)

And we proceeded to demonstrate that numerous leading neuroscientists could indeed be condemned out of their own mouth, precisely because they draw inferences from their application of

the psychological vocabulary to the brain that can only intelligibly be drawn from its customary application to the animal as
a whole (PFN chapters 3–8).

If a neuroscientist applies psychological expressions or semantic expressions such as "representations" or expressions like
"map" to the brain, then he is either using these in their customary sense or he is not. If the latter, then his use may be or
involve 1. a derivative sense, 2. an analogical or other extension
of the old term, 3. a mere homonym, or 4. a metaphorical or
figurative sense. If psychological terms are applied to the brain
in their customary sense, then what is said is not intelligible. We
do not know what it means to say that the brain thinks, fears,
or is ashamed. The constitutive grounds upon which competent
speakers of our language apply such expressions to animals and
human beings, namely, what they say and do, cannot be satisfied
by a brain or its parts—there is no such thing as a brain or part
of a brain making thoughtful observations, running away in fear,
or blushing in shame. We no more understand what it would be
for a brain or its parts to think, reason, fear, or decide something
than we understand what it would be for a tree to do so. If such
terms are being applied in a novel sense, then the user owes us
an explanation of what that sense is. It may be a derivative sense,
as when we apply the term "healthy" to food or exercise—a use
that needs a different explanation from the explanation appropriate for the primary use of "health" in application to a living
being. It may be an analogical use, as when we speak of the foot
of a mountain or of a page—such analogies are typically evident,
but obviously call out for a very different paraphrastic explanation than that demanded by their prototype. Or it may be a
homonym, like "mass" in Newtonian mechanics, which requires
a quite different explanation from "mass" in "a mass of people"
or "a mass of poppies."

Neuroscientists' use of "representation" is, for the most part, intended as a mere homonym of "representation" in its symbolic and
semantic sense. This has turned out to be ill-advised, for eminent

scientists and psychologists have succumbed to the confusion of using the word both to mean a causal correlate or a concomitance and also to mean a symbolic representation. For it is in the former sense alone that it makes sense to speak of representations in the brain. Hence our criticisms of David Marr (PFN 70, 76, 143–47). Neuroscientists' use of the term "map" appears to have begun life as an extension of the idea of a mapping, but it rapidly became confused with that of a map. There is nothing wrong with calling the set of entities onto which members of another set can be mapped "a map" of the latter—although it is neither necessary nor clear. But incoherence is afoot if one then suppose that this "map" might be used by the brain in the manner in which readers of an atlas use maps. It is altogether obscure what is meant by the claim that the brain or its parts know, believe, think, infer, and perceive things. The only coherent idea that might be lurking here is that these terms are applied to the brain to signify the neural activity that supposedly corresponds with the animal's knowing, believing, thinking, inferring, and perceiving. But then one cannot intelligibly go on to assert (as Crick, Sperry, and Gazzaniga do) that the part of the brain that is thinking communicates *what it thinks* to another part of the brain. For, while human thinking has a content (given by the answer to the question "What are you thinking?"), neural activity cannot be said to have any content whatsoever.

It might be suggested that neuroscientists' talk of *maps* or *symbolic descriptions* in the brain and of the brain's knowing, thinking, deciding, interpreting, etc. is metaphorical.[28] These terms, one might claim, are actually probing metaphors the aptness of which is already long established with regard to *electronic computers*, which are aptly described as "following rules." For computers were "*deliberately built* to engage in the 'rule-governed manipulation of complex symbols.'" Indeed, one might think, such talk "is not even metaphorical any longer, given the well developed theoretical and technological background against which such talk takes place."[29] Similarly, cognitive neuroscientists, in their use of the common psychological vocabulary "are indeed groping forward in the darkness; metaphors are indeed the rule rather than the exception." But this is normal scientific progress, and in some cases, neuroscience has advanced beyond metaphor, e.g., in ascription of "sentence-like representations" and "map-like representations" to the brain.

This can be questioned. Computers cannot correctly be described as following rules any more than planets can correctly be described as complying with laws. The orbital motion of the planets is *described* by the Keplerian laws, but the planets do not *comply* with the laws. Computers were not built to "engage in rule-governed manipulation of symbols," they were built to produce results that will *coincide* with rule-governed, *correct* manipulation of symbols. For computers can no more *follow* a rule than a mechanical calculator can. A machine can execute operations that accord with a rule, provided all the causal links built into it function as designed and assuming that the design ensures the generation of a regularity in accordance with the chosen rule or rules. But for something to constitute following a rule, the mere production of a regularity in accordance with a rule is not sufficient. A being can be said to be following a rule only in the context of a complex practice involving actual and potential activities of justifying, noticing mistakes and correcting them by reference to the rule, criticizing deviations from the rule, and, if called upon, explaining an action as being in accordance with the rule and teaching others what counts as following a rule. The determination of an act as being *correct*, in accordance with the rule, is not a causal determination but a logical one. Otherwise we should have to surrender to whatever results our computers produce.[30]

To be sure, computer engineers use such language—harmlessly, until such time as they start to treat it literally, and suppose that computers really think, better and faster than we do, that they truly remember, and, unlike us, never forget, that they interpret what we type in, and sometimes misinterpret it, taking what we wrote to mean something other than we meant. *Then* the engineers' otherwise harmless style of speech ceases to be an amusing shorthand and becomes a potentially pernicious conceptual confusion.

To say that computers or brains think, calculate, reason, infer, and hypothesize *may* be intended metaphorically. Metaphors do not explain—they illustrate one thing in terms of another. A metaphor ceases to be a metaphor when it becomes a dead metaphor, like "a broken heart" or "at one fell swoop." But it cannot cease to be a metaphor by becoming literal. What would it be for it to be *literally* true that the planets obey the laws of nature? A slide rule, mechanical calculator, or computer can be said to calculate—figuratively speaking. But what would it be for it to do so literally? If "the computer remembers (calculates, infers, etc.)" makes perfect (nonfigurative) sense to computer engineers, that is precisely because they treat these phrases as dead metaphors. "The computer calculates" means no more than "The computer goes through the electricomechanical processes necessary to produce the results of a calculation without any calculation," just as "I love you with all my heart" means no more than "I truly love you."

It is noteworthy that neuroscientists' talk of there being maps in the brain, and of the brain using these maps as maps, is, in the cases that we criticized, anything but metaphorical. Colin Blakemore's remark that "neuroanatomists have come to speak of the brain having *maps*, which are thought to play an essential part in the representation and interpretation of the world by the brain, *just as the maps of an atlas do for the reader of them*"[31] (emphasis added) is obviously not metaphorical, since there is nothing metaphorical about "Maps of an atlas play a role in the representation of the world for their readers." Moreover, "representation" here is patently used in the symbolic sense, not the causal correlate sense. Nor is J. Z. Young's assertion that the brain makes use of its maps in formulating its hypotheses about what is visible.[32] For *to make use of a map in formulating a hypothesis* is to take a feature indicated by the map as a *reason* for the hypothesis. Professor Dennett asserted that the brain "*does* make use of them *as* maps" (p. 205, n. 20), and in the debate at the APA he asserted that it is an empirical, not a philosophical, question whether "retinotopical maps" are used by the brain *as maps*, "whether any of the information-retrieval operations that are defined over them exploit features of maps that we exploit when we exploit maps in the regular world." But it can be an empirical question only if it made sense for the brain to use a map as a map. However, to *use* a map as a map, there has to *be* a map—and there are none in the brain; one has to be able to read the map—but brains lack eyes and cannot read; one has to be familiar with the projective conventions of the map (e.g., cylindrical, conic, azimuthal)—but there are no projective *conventions* regarding the mappings of features of the visual field onto the neural firings in the "visual" striate cortex; and one has to use the map to guide one's behavior—one's perambulations or navigations—which are not activities brains engage in. One must not confuse a map with the possibility of a mapping. That one can map the firing of retinal cells onto the firing of cells in the visual striate cortex does not show that there is a map of visibilia in the visual field in the visual striate cortex.

Finally, we should like to rectify a misunderstanding. Some of our critics assume that we are trying to lay down a law prohibiting novel extensions of expressions in the language. Professor Dennett asserted in the debate at the APA that we would outlaw talk of the genetic *code*, given that we insist that knowledge cannot be *encoded* in the brain. Professor Churchland supposes that we would in principle exclude such conceptual innovations as Newton's talk of the moon constantly *falling* toward the earth as it moves upon its inertial path. This is a misunderstanding.

We are not prohibiting anything—only pointing out when conceptual incoherences occur in neuroscientific writings. We are not trying to stop anyone from extending usage in scientifically fruitful ways—only trying to ensure that such putative extensions do not transgress the bounds of sense through failure adequately to specify the novel use or through crossing the novel use with the old one. There is nothing wrong with talking about the *foot* of a mountain—as long as one does not wonder whether it has a shoe. There is nothing wrong with speaking of *unhealthy* food—as long as one does not wonder when it will regain its health. There is nothing wrong with talk of the *passing* of time—as long as one does not get confused (as Augustine famously did) about how to measure it. There was nothing wrong with Newton's talking about the moon's "falling"—but there would have been had he wondered what made the moon slip. There was nothing wrong with his speaking of forces acting on a body in space—but there would have been had he speculated whether the forces were infantry or cavalry. There is nothing wrong with geneticists speaking of the genetic code. But there would be if they drew inferences from the existence of the genetic code that can be drawn only from the existence of literal codes. For, to be sure, the genetic code is not a code in the sense in which one uses a code to encrypt or transmit a sentence of a language. It is not even a code in "an attenuated sense," as a sentence agreed between husband and wife to talk over the heads of the children might be deemed to be.

Our concern was with the use, by cognitive neuroscientists, of the common or garden psychological vocabulary (and other terms such as "representation" and "map") in specifying the explananda of their theories and in describing the explanans. For, as we made clear, neuroscientists commonly try to explain human beings' perceiving, knowing, believing, remembering, deciding by reference to parts of the brain perceiving, knowing, believing, remembering, and deciding. So we noted such remarks, made by leading neuroscientists, psychologists, and cognitive scientists, as the following:

> J. Z. Young: "We can regard all seeing as a continual search for the answers to questions posed by the brain. The signals from the retina constitute 'messages' conveying these answers. The brain then uses this information to construct a suitable hypothesis of what is there."
>
> *Programs of the Brain*, p. 119

C. Blakemore: "The brain [has] maps, which are thought to play an essential part in the representation and interpretation of the world by the brain, just as the maps of an atlas do for the readers of them."

"Understanding Images in the Brain," p. 265

G. Edelman: The brain "recursively relates semantic to phonological sequences and then generates their syntactic correspondences . . . by treating rules developing in memory as objects for conceptual manipulation."

Bright Air, Brilliant Fire (Harmondsworth: Penguin, 1994), p. 130

J. Frisby: "There must be a symbolic description in the brain of the outside world, a description cast in symbols which stand for various aspects of the world of which sight makes us aware."

Seeing: Illusion, Brain, and Mind
(Oxford: Oxford University Press, 1980), p. 8

F. Crick: "When the callosum is cut, the left hemisphere sees only the right half of the visual field . . . both hemispheres can hear what is being said . . . one half of the brain appears to be almost totally ignorant of what the other half saw."

The Astonishing Hypothesis (London: Touchstone, 1995), p. 170

S. Zeki: "The brain's capacity to acquire knowledge, to abstract and to construct ideals."

"Splendours and Miseries of the Brain," *Philosophical Transactions of the Royal Society B* 354 (1999), p. 2054

D. Marr: "Our brains must somehow be capable of representing . . . information. . . . The study of vision must therefore include . . . an inquiry into the nature of the internal representations by which we capture this information and make it available as a basis for decisions . . .

A representation is a formal scheme for describing ... together with rules that specify how the scheme is to be applied. ... [A formal scheme is] a set of symbols with rules for
putting them together. ... A representation, therefore, is not
a foreign idea at all—we all use representations all the time."

Vision (San Francisco: Freeman, 1980), pp. 20f.

These are not metaphorical uses. They are not bold extensions of
terms, introducing new meanings for theoretical purposes. They
are simply misuses of the common psychological (and semantic)
vocabulary—misuses that lead to incoherence and various forms
of nonsense—that we pointed out from case to case. There is
nothing surprising about this. It is no different, in principle, from
the equally misguided applications of the same vocabulary *to the
mind*—as if it were my mind that knows, believes, thinks, perceives, feels pain, wants, and decides. But it is not; it is I, the living human being that I am, that does so. The former error is no
less egregious than the (venerable) latter one, and is rife among
cognitive neuroscientists—sometimes to the detriment of the
experiments they devise, commonly to their theorizing about
the results of their experiments, and often to their explaining
animal and human cognitive functions by reference to the neural
structures and operations that make them possible.

Qualia

In our discussion of consciousness (PFN chapters 9–12) we argued that characterizing the domain of the mental by reference
to the "qualitative feel" of experience is misconceived (PFN
chapter 10). But *pace* Professor Searle (p. 99f.), we did not deny
the existence of qualia on the grounds that if they did exist they
would exist in brains. If, *per impossibile,* psychological attributes
were all characterized by their "qualitative feel," they would still
be attributes of human beings, not of brains.

A quale is supposed to be "the qualitative feel of an experience" (Chalmers),[33] or it is such a thing as "the redness of red or
the painfulness of pain" (Crick).[34] Qualia are "the simple sensory
qualities to be found in the blueness of the sky or the sound of
a tone" (Damasio)[35] or "ways it feels to see, hear and smell, the
way it feels to have a pain" (Block).[36] According to Professor
Searle, conscious states are "qualitative in the sense that for any
conscious state . . . there is something that it qualitatively feels
like to be in that state."[37] According to Nagel, for every conscious experience "there is something it is like for the organism
to have it."[38] These various explanations *do not amount to the same
thing*, and it is questionable whether a coherent account emerges
from them.

Professor Searle remarks that there is a qualitative feel to a
pain, a tickle, and an itch. To this we agree—in the following
sense: sensations, we remarked (PFN 124), have phenomenal
qualities (e.g., burning, stinging, gnawing, piercing, throbbing);
they are linked with felt inclinations to behave (to scratch, assuage, giggle, or laugh); they have degrees of intensity that may
wax or wane.

When it comes to perceiving, however, we noted that it is
problematic to characterize what is meant by "the qualitative
character of experience." Specifying what we see or smell, or, in
the case of hallucinations, what it seems to us that we see or smell,
requires specification of an object. Visual or olfactory experiences and their hallucinatory counterparts are individuated by
what they are experiences or hallucinations of. Seeing a lamppost
is distinct from seeing a mailbox, smelling lilac is different from
smelling roses, and so too are the corresponding hallucinatory
experiences that are described in terms of their seeming to the
subject to be like their veridical perceptual counterpart.[39]

To be sure, roses do not smell like lilac—what roses smell like
is different from what lilac smells like. Smelling roses is quite different from smelling lilac. But the qualitative character of smelling roses does not smell of roses, any more than the qualitative

character of smelling lilac smells of lilac. Smelling either may be equally pleasant—in which case the qualitative character of the smelling may be exactly the same, even though what is smelled is quite different. Professor Searle, we suggest, confuses what the smells are like with what the smelling is like.

Seeing a lamppost does not normally feel like anything. If asked "What did it feel like to see it?" the only kind of answer is one such as "It didn't feel like anything in particular—neither pleasant nor unpleasant, neither exciting nor dull." Such epithets—"pleasant," "unpleasant," "exciting," "dull"—are correctly understood as describing the "qualitative character of the experience." In this sense, *many* perceptual experiences have no qualitative character at all. *None* are individuated by their qualitative feel—they are individuated by their object. And if we are dealing with a hallucination, then saying that the hallucinated lamppost was black is still the description of the object of the experience—its "intentional object" in Brentano's jargon (which Professor Searle uses). The quality of the hallucinatory experience, on the other hand, is probably: *rather scary*.

Contrary to what Professor Searle suggests, we did not argue that "if you do not define qualia as a matter of pleasantness or unpleasantness, then you will have to individuate the experience by its object" (p. 115). Our argument was that we *do* individuate experiences and hallucinations by their objects—which are specified by the answer to the question "What was your experience (or hallucination) an experience (or hallucination) *of*?"[40] Of course, the object need not be the cause, as is evident in the case of hallucinations. But, we insisted, the qualitative character of the experience should not be confused with the qualities of the object of the experience. That what one sees when one sees a red apple is red and round does not imply that one had a red, round visual experience. That what one seems to see when one hallucinates a red apple is red and round does not imply that one had a red, round visual hallucination. "What did you see (or hallucinate)?" is one question, "What was it like to see what you saw (or hallucinate what you hallucinated)?" another. One does not individuate perceptual experiences by their qualitative character. These are simple truths, but they seem to have been overlooked.

Enskulled Brains

Professor Searle suggests that human beings are "embodied brains" (pp. 120f.). According to his view, the reason why we can say both "I weigh 160 lbs" and "My body weighs 160 lbs" is that what makes it the case that I weigh 160 lbs is that my body does. But I, it seems, am strictly speaking no more than an embodied (enskulled) brain. I *have* a body, and I am *in* the skull *of* my body. This is a materialist version of Cartesianism. One major reason why we wrote our book was the firm belief that contemporary neuroscientists, and many philosophers too, still stand in the long, dark shadow of Descartes. For while rejecting the immaterial substance of the Cartesian mind, they transfer the attributes of the Cartesian mind to the human brain instead, leaving intact the whole misconceived structure of the Cartesian conception of the relationship between mind and body. What we were advocating was that neuroscientists, and even philosophers, leave the Cartesian shadow lands and seek out the Aristotelian sunlight, where one can see so much better.

If I were, *per impossibile*, an embodied brain, then I would have a body—just as the Cartesian embodied mind has a body. But I would not *have* a brain, since brains do not have brains. And in truth my body would not weigh 160 lbs, but 160 lbs less 3 lbs— which is, strictly speaking, what *I* would weigh. And I would not be 6 foot tall, but only 7 inches tall. Doubtless Professor Searle will assure me that I am my-embodied-brain—my brain *together* with my body. But that does not get us back on track. For my brain together with my brainless body, taken one way, is just my cadaver; taken another way, it is simply *my body*. But I am not my body, not the body *I have*. Of course, I am *a* body—the living human being that stands before you, a particular kind of sentient spatiotemporal continuant that possesses intellect and will and is therefore a person. But I am no more *my body* than I am *my mind*. And I am not an embodied brain either. It is mistaken to suppose that human beings are "embodied" at all—that conception belongs to the Platonic, Augustinian, and Cartesian tradition that should be repudiated. It would be far better to say, with Aristotle, that human beings are *ensouled* creatures (*empsuchos*)—animals endowed with such capacities that confer upon them, in the form of life that is natural to them, the status of persons.

Neuroscientific Research

Our critics suggest that our investigations are irrelevant to neuroscience or—even worse—that our advice would be positively harmful if followed. Professor Dennett holds that our refusal to ascribe psychological attributes (even in an attenuated sense) to anything less than an animal as a whole is retrograde and unscientific. This, he believes, stands in contrast to the scientific benefits of the "intentional stance" that he advocates. In his view, "the poetic license granted by the intentional stance eases the task" of explaining how the functioning of parts contributes to the behavior of the animal (p. 89).

We note first that *poetic license* is something granted to poets for purposes of poetry, not for purposes of empirical precision and explanatory power. Second, ascribing cognitive powers to parts of the brain provides only the semblance of an explanation where an explanation is still wanting. So it actually blocks scientific progress. Sperry and Gazzaniga claim that, in cases of commissurotomy, the bizarre behavior of subjects under experimental conditions of exposure to pictured objects is *explained* by the fact that one hemisphere of the brain is ignorant of what the other half can see. The hemispheres of the brain allegedly know things and can explain things, and, because of the severance of the corpus callosum, the right hemisphere allegedly cannot communicate to the left hemisphere what it sees. So the left hemisphere must generate its own interpretation of why the left hand is doing what it is doing.[41] Far from explaining the phenomena, this masks the absence of any substantial explanation by redescribing them in misleading terms. The dissociation of functions normally associated is indeed partially explained by the severing of the corpus callosum and by the localization of function in the two hemispheres. *That* is now well known, but currently available explanation goes no further. It is an illusion to suppose that anything whatsoever is added by ascribing knowledge, perception, and linguistic understanding ("sort of" or otherwise) to the hemispheres of the brain.

Professor Searle claims that central questions in neurobiological research would be rejected as meaningless if our account of the conceptual structures deployed were correct. So, he suggests, "the central question in vision, how do neurobiological processes . . . cause conscious visual experiences, could not be investigated by anyone who accepted [our] conception." Our conception, he avers, "can have potentially disastrous scientific consequences" (p. 124).

Research on the neurobiology of vision is research into the neural structures that are causally necessary for an animal to be able to see and into the specific processes involved in its seeing. That we deny that visual experiences occur in the brain, or that they are characterized by qualia, affects this neuroscientific research program only insofar as it averts futile questions that could have no answer. We gave numerous examples, e.g., the binding problem (Crick, Kandel, and Wurtz) or the explanation of recognition by reference to the matching of templates with images (Marr) or the suggestion that perceptions are hypotheses of the brain that are conclusions of unconscious inferences it makes (Helmholtz, Gregory, and Blakemore). Our contention that it is the animal that sees or has visual experiences, not the brain, and Professor Searle's contention that it is the brain, not the animal, are conceptual claims, not empirical ones. The issue is nonetheless important for all that, but it should be evident that what we said does not *hinder* empirical investigation into the neural processes that underpin vision. Rather, it guides the description of the results of such investigations down the highroads of sense.

In general, the conceptual criticisms in our book do no more than peel away layers of conceptual confusion from neuroscientific research and clarify the conceptual forms it presupposes. This cannot impede the progress of neuroscience. Indeed, it should facilitate it—by excluding nonsensical questions, preventing misconceived experiments, and reducing misunderstood experimental results.[42]

EPILOGUE

I met Sir John Eccles in 1962 while concluding my degree in electrical engineering. The following year he was awarded the Nobel Prize for his work on chemical transmission at synapses in the spinal cord and brain. He asked me what I was doing and I replied "electrical engineering," to which he responded, "Excellent, you should join me, as every first-rate neurophysiology laboratory needs a very good solderer." I believe that every first-rate cognitive neuroscience laboratory now needs a very good critical, analytical philosopher. The dialogue concerning the aims and accomplishments of cognitive neuroscience at the 2005 American Philosophical Association meeting in New York, detailed in this book, supports my case.

Neuroscience is concerned with understanding the workings of the nervous system, thereby helping in the design of strategies to relieve humanity of the dreadful burden of such diseases as dementia and schizophrenia. Neuroscientists, fulfilling this task, also illuminate those mechanisms in the brain that must function normally for us to be able to exercise our psychological faculties, such as perception and memory.

This view of neuroscience is opposed to one that holds that neuroscience has a single overarching goal, namely, to understand *consciousness*.[1] It is interesting to consider this suggestion at some length as an example of the need for critical philosophical analysis of the kind I have called for in the neurosciences. Peter Hacker and I allocated over one hundred pages to the subject of *consciousness* in our book. At the beginning of our analysis we state that "a first step towards clarity is to distinguish *transitive* from *intransitive* consciousness. Transitive consciousness is a matter of being conscious *of* something or other, or of being conscious *that* something or other is thus or otherwise. Intransitive consciousness, by contrast, has no object. It is a matter of being conscious or awake, as opposed to being unconscious or asleep" (*Philosophical Foundations of Neuroscience*, hereafter PFN, p. 244). The loss of intransitive consciousness, as in sleeping, fainting, or being anesthetized, is the subject of a rich neuroscientific literature. On the other hand, there is much confusion in the neuroscientific literature when it comes to the study of the various forms of transitive consciousness. These include perceptual, somatic, kinesthetic, and affective consciousness, consciousness of one's motives, reflective consciousness, consciousness of one's actions, and self-consciousness (PFN, pp. 248–52). Some of these forms of transitive consciousness are attentional. For example, perceptual consciousness of something involves having one's attention caught and held by the thing of which one is conscious. It is mistaken to suppose that perceiving something is, as such, a form of transitive consciousness or even entails being conscious of what one perceives.

Neuroscientific research has been devoted overwhelmingly to only one form of what is conceived (or misconceived) to be transitive perceptual consciousness, namely, that involved in visual perception, particularly the phenomenon of binocular rivalry.

During such rivalry, the observer views two incongruent images, one affecting each eye, but perceives only one image at a time. The image that is perceptually dominant alternates every few seconds. This can be understood by means of the experimental work of Logothetis and his colleagues (Leopold and Logothetis 1999; Blake and Logothetis 2002). Using operant conditioning techniques, monkeys are trained to operate a lever to indicate which of the two competing monocular images is dominant at any given time (see figure 8; left-hand side, upper panel). Action potential spike activity, recorded from single cells in the monkey's visual cortex, can be correlated with the animal's perceptual response of moving the lever. This makes it possible to identify the cortical regions in which neuronal activity corresponds with perceptual experience. The lower panel (left-hand side of figure 8) shows the number of spikes recorded from a single active neuron during binocular rivalry. The bar along the x axis indicates alternating perception of the two images that are clearly correlated with the periods of spike firing. Figure 8 (right-hand side) shows the brain areas that contained responsive neurons whose activity correlated with the monkey's visual perception. The percentage of perception-related neurons increases in the "higher" visual centers, that is, those furthest removed from the input to the cortex from the thalamus. Only a small fraction of neurons responsive to visual stimuli in the earliest cortical areas V1 (with direct connection to the thalamus) and V2 responded in concert with the binocular rivalry alternations, whereas the percentage was higher in areas furthest removed from the thalamic input, namely, V4, MT, and MST. The activity of nearly all visually responsive neurons in areas IT and STS closely matched the animal's perceptual state. This modulation of cortical neurons contrasts with the activity in noncortical neurons of the lateral geniculate nucelus that have direct connections with the retina. These do not show any modulation during binocular rivalry.

There is thus a distributed set of neurons in the cortex that fire in coordination with the animal's perceptual responses, albeit

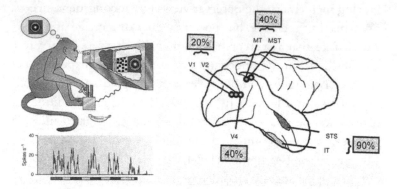

FIGURE 8. Operant conditioning techniques to establish the areas of cortex containing neurons that fire action potential spikes correlated with the monkey's "perceptual reports" during presentation of "competing" monocular images (from Blake and Logothetis 2002; Leopold and Logothetis 1999).

less neurons in the lower visual areas (V1) than in the higher ones. The spatial distribution of neurons in the cortex that are active during such perceptual responses to binocular alternation has been emphasized by Lumer and his colleagues, using functional magnetic resonance imaging on human subjects. Their research revealed a larger distribution over the cortex of active neurons whose activity fluctuates coordinate with fluctuations in the reported perceptual experience, taking in the lateral prefrontal cortex as well as the higher visual centres (Lumer, Friston, and Rees 1998). This has also been reported by Edelman and his colleagues using neuromagnetic recordings from the cortex of human subjects (Tononi and Edelman 1998; Srinivasan et al. 1999). This kind of distributed activity during perception appears to be what has led Professor Searle to suggest that there is a "consciousness field."[2]

By contrast with emphasizing the distributed activity in the cortex during binocular rivalry, others suggest that specific classes of neurons in higher visual centers are alone to be considered as providing the NCC for this phenomenon. For example, Crick and Koch (2003) stress that it is only the higher visual centers, like STS and IT in figure 8, which possess visually activated neurons that typically respond (90 percent) in concert with the binocular rivalry alternations. This leads Crick and Koch to examine the details of "the dendritic aborizations of the different types of neurons in the inferior temporal (IT) gyrus of the macaque monkey that project to the prefrontal cortex near the principle sulcus" (figure 9; top, shaded grey). They go on to "note that only one type of cell has apical dendrites that reach to layer 1." They then ask, "What could be special about the activity that reaches above the consciousness threshold? It might be the firing of special types of neurons, such as those pyramidal cells that project to the front of the brain" (figure 9). So we now have the NCC of the perception during binocular rivalry identified with a particular neuron type in the IT region of cortex, an idea akin to that of the "pontifical cell" or "cardinal cells" of Barlow (1997). This leads Koch (2004) in his book *The Quest for Consciousness*, greatly admired by some senior philosophers,[3] to describe what neuroscience might have revealed concerning the NCC during visual perception by means of figure 10. This is a crypto-Cartesian view of the relation between our psychological attributes and the workings of the brain. It is rooted in the misconception, which we assailed in our book (PFN, chapter 10), that the essence of consciousness is its association with qualia and hence that these ineffable qualitative characteristics of experience might be found to be caused by cardinal or pontifical cells.

I believe that this brief description of neuroscientific research and its interpretation concerning visual perceptions during binocular rivalry reveals an urgent need for critical clarification from analytical philosophers. Peter Hacker and I suggest that the results of this research, whether interpreted in terms of a "consciousness field" or of "cardinal cells," do not contribute to an understanding of transitive perceptual consciousness at all. At most they contribute to the identification of some of the

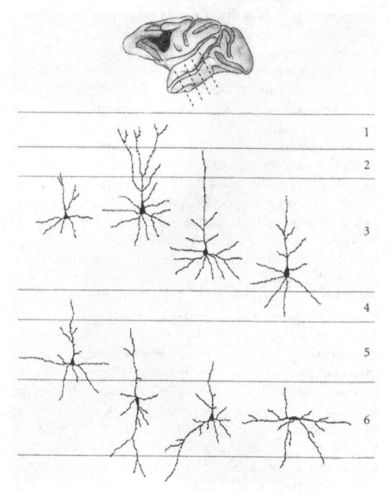

FIGURE 9. A collage of neurons in the inferior-temporal cortex (IT in figure 8) that project to a limited portion of the prefrontal cortex (stippled area in the insert brain at the top; from Crick and Koch 2003, derived from De Lima, Voigt, and Morrison 1990).

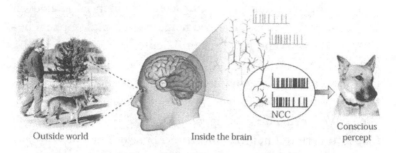

Outside world   Inside the brain   Conscious percept

FIGURE 10. "The NCC are the minimal set of neural events—here synchronized action potentials in neocortical pyramidal neurons (in area IT)—sufficient for a specific conscious percept" (from Koch 2004).

neural correlates of visual perception under conditions of binocular rivalry. But transitive perceptual consciousness involves having one's attention caught and held by something in the visual field. The neuroscientific study that is required to discover the NCC of transitive perceptual consciousness is into the neural correlates of having one's attention caught and held by what one perceives, not into perception per se. For to perceive an object is not the same as being conscious of the object one perceives. One may perceive an object X without being conscious of it, either because one misidentifies it as Y or because one's attention is not caught and held by it, which may be because one does not even notice it or because one is *intentionally* attending to it. Transitive perceptual consciousness is a form of cognitive *receptivity* (PFN, pp. 253–60). Affective consciousness, consciousness of one's motives, reflective consciousness of one's actions, and self-consciousness require importantly different forms of analysis. Many other examples indicating the need for clarification, and covering the whole range of neuroscientific inquiries into the neuronal correlates of our psychological attributes, are given in *Philosophical Foundations of Neuroscience* and in our forthcoming book, *History of Cognitive Neuroscience: A Conceptual Analysis*.

The misplaced hubris in neuroscience, which I touched on in the introduction, will only be heightened if philosophers become acolytes to the neuroscientific enterprise. Writing laudatory reviews in the *New York Review of Books* on the works of neuroscientists, who seem unaware of the patent conceptual difficulties associated with their ideas, are not what the discipline needs. We need illuminating philosophical criticism that will help guide fruitful neuroscientific research into our psychological powers and their exercise. This, I believe, is a major task for the younger generation of philosophers.

References

Barlow, H. 1997. The neuron doctrine in perception. In M. S. Gazzaniga, ed., *The New Cognitive Neurosciences*, 4th ed. (Cambridge: MIT Press), p. 421.

Blake, R., and N. K. Logothetis. 2002. Visual competition. *Nature Reviews Neuroscience* 3:13–21.

Crick, F., and C. Koch. 2003. A framework for consciousness. *Nature Reviews Neuroscience* 6:119–26.

De Lima, A. D., T. Voigt, and J. H. Morrison. 1990. Morphology of the cells within the inferior temporal gyrus that project to the prefrontal cortex in the macaque monkey. *Journal of Comparative Neurology* 296:159–72.

Koch, C. 2004. *The Quest for Consciousness* (Greenwood Village, CO: Roberts).

Leopold, D. A., and N. K. Logothetis. 1999. Multistable phenomena: changing views in perception. *Trends in Cognitive Science* 3:254–64.

Lumer, E. D., K. J. Friston, and G. Rees. 1998. Neural correlates of perceptual rivalry in the human brain. *Science* 280:1930–34.

Srinivasan, R., D. P. Russell, G. M. Edelman, and G. Tononi. 1999. Increased synchronization of neuromagnetic responses during conscious perception. *Journal of Neuroscience* 19:5435–48.

Tononi, G., and G. M. Edelman. 1998. Consciousness and complexity. *Science* 282:1846–51.

STILL LOOKING

Science and Philosophy in Pursuit of Prince Reason

DANIEL ROBINSON

Anatomy was in the air of the thinking classes in the Britain of the early seventeenth century. The great William Harvey, returning from Italy in 1602, would offer his pioneering Lumleian lectures from 1615. Cambridge was one of the centers of this revived interest in the machinery of the body. Harvey, degree in hand, left there in 1600 to receive instruction in Padua from Fabricius himself, even as the younger Phineas Fletcher (1582–1650) was completing his own course of study at King's College. We all know about Harvey. Fletcher is nearly lost in the mists of time. Nor would any student of anatomy trade Fletcher's mode of instruction for Harvey's. However, as the manner in which we comprehend all or any part of the natural world is rigidly bound to the methods selected for the task, curiosity is repaid by reviving Fletcher's approach.

Published in 1633, Phineas Fletcher's *The Purple Island* is an allegory in twelve cantos guiding the reader into and all through the mysterious terrain that is the human body. The island owes its hue to the purple stuff out of which God fashioned the new earth. The song of anatomical discovery is sung by Thirsil to an audience of young shepherds. If we think of these swains as similar to today's students, we might agree that it is not until Thirsil reaches canto 6 that interest is piqued, for in stanza 28 we enter the realm of

> The Islands Prince, of frame more than celestiall,
> ... rightly call'd th'all-seeing *Intellect*;
> All glorious bright, such nothing is terrestriall;
> Whose Sun-like face, and most divine aspect
>   No humane sight may ever hope descrie:
>   For when himself on's self reflects his eye,
> Dull or amaz'd he stands at so bright majestie.

Then, proceeding to stanza 30, no doubt is left as to the composition of the Prince:

> His strangest body is not bodily,
> But matter without matter; never fill'd,
> Nor filling; though within his compasse high
> All heav'n and earth, and all in both are held;
>   Yet thousand thousand heav'ns he could contain,
>   And still as empty as at first remain;
> And when he takes in most, readi'st to take again.[1]

Harvey and Fletcher had joined the long search for the seat of the rational soul, for that "place of forms" Aristotle wisely did not seek to locate, that "strangest body" whose defining properties seem anything but bodily. As the present volume makes clear, we are still looking. As the present volume makes even clearer, however, there is now less certainty about just what it is one is likely to find in the more likely places.

Contributors to this volume include a prominent scientist and a group of highly accomplished and influential philosophers.

If the differences that animate their exchanges are to be located within the larger context of intellectual history, it is useful to consider once more the methods of Harvey and of Fletcher, each setting out on a voyage of discovery, each beholden to a mode of explanation seemingly vindicated by human practices of proven worth. Understood in this light, and with reservations duly noted later, I would suggest that John Searle and Daniel Dennett would wish to be identified with Harvey, even as their speculative philosophies are actually in the patrimony of Fletcher. As with Fletcher, they are students of the anatomy of their time but would use it to tell a story. It is the reader's option to treat the results either as allegory or as headline news. Insofar as it is a story, however, it is not to be mistaken for the quite different mission of either experimental or theoretical science.

Max Bennett and Peter Hacker, in skillfully drawing attention to this, would press on to insist that it is in the very nature of the case, that these are distinct paths leading toward worthy but fundamentally different goals. Passing no judgment here on the relative value of distinct modes of inquiry and of explanation, one might note that the pages contributed by Bennett and Hacker (excepting Bennett's explicitly technical pages) are rooted in the long-accepted traditions of analytical philosophy. I refer here not to some alleged "discovery" by a band of plain-thinking Oxford philosophers but to that central mission of Plato's dialogues— the clarification of terms, the casting of problems in argumentative form, the demand for consistency and coherence. To all this Aristotle would add the content of the natural world and the greatly enlarged philosophical mission arising from that addition. Dennett and Searle have stories to tell, and they are good stories told by masters of the craft. Bennett and Hacker, in their significant book, *Philosophical Foundations of Neuroscience*, concluded that the truth-value of such stories could not be assessed owing to the peculiar, unscientific, and unphilosophical choice of terms. What they found in Searle, Dennett, and other members of today's leaders of thought in cognitive neuroscience was just too much of Gilbert Ryle's "she arrived in a veil of tears and a sedan chair." In the present volume, readers are drawn back into the stories in a condensed and instructively dialectical form.

In reflecting on his own chapter, Bennett recounts a meeting with John Eccles who teasingly insisted that research in neurophysiology always requires "a very good solderer." Bennett then expresses his own conviction that "every first-rate cognitive neuroscience laboratory now needs a very good critical, analytical philosopher." Having devoted many years to both the bench and the armchair, I know firsthand that Eccles was right, though I am less convinced than Bennett is. My doubt springs from a more pervasive scepticism in the matter of compounding or hyphenating well-defined disciplines. Once the well-defined discipline of ethics is reworked into something called bioethics, there is a tendency to think that some deeper ethical precept must be found to cover instances of stealing someone's liver, the precepts covering auto theft being insufficient unto the task. With "cognitive neuroscience," the very adjective seemingly settles an issue that has been philosophically refractory at least since the days when Plato gave voice to the Socrates of the *Crito*. Bennett's early wonder—How, indeed, to derive the psychological from the synaptic!—is Fletcher's wonder. What Fletcher lacked, as he attempted to *see* the Prince, could not have been supplied by Harvey. Neither the dissecting needle nor the anatomical blowpipe could be serviceable here. Strong arguments remain to the effect that the larger mission of a realistic and informed *cognitive psychology* must begin along a road different from that which guides the progress of the scientist as pilgrim.

Perhaps I can make my point sharper by referring to the very research summarized by Bennett to establish how visual perceptual outcomes are correlated more strongly with "higher" cortical events than with those closer to their retinal origins. There is a long and consistent series of findings establishing that, at least in audition and vision, the "tuning" of the system becomes ever sharper as events move from the level of the first-order neurons to their ultimate cortical destinations. But the phenomenon of binocular rivalry is different from the tuning of the system to narrower bands of frequencies. There is a seeming phenomenological property signaled by the animal's operant behavior as first one and then the other image becomes dominant. The question that naturally arises within this framework has to do with the manner in which all this takes place in creatures with laterally placed eyes, such that there cannot be the same type of binocular *rivalry*. Animals with medially placed eyes confront a visual space in which the same object might compete for recognition. Those with laterally placed eyes confront two separated visual spaces that will have no object in common.

Why do I mention this? I do so in order to make the obvious (if routinely overlooked) point that there is not only an actual *animal* that is seeing something, but that the manner in which the visible environment is thus engaged will depend on a far wider and more complex ecological reality with which creatures of a given kind must come to terms. This fact imposes limits of varying degrees of severity on generalizations across species. It is permissible to suspect that even tighter limits are actually at work when generalizations include either "transitive" or "intransitive" consciousness. It is in the sense that fish will not discover water that observations within the bubble of the laboratory seem hopelessly remote from life lived in the visible world. Put another way, it would seem to be far easier to settle on what it is like to be a cat than what is like to have seen nothing beyond what can be projected on to disparate retinal loci during a lifetime spent in a laboratory cage.

Especially interesting among the cautions Bennett and Hacker announce to otherwise unsuspecting cognitive neuroscientists is one that has to do with the claim that we are not always conscious of what we perceive. Bennett notes that "one may perceive an object X without being conscious of it, either because one misidentifies it as Y, or because one's attention is not caught and held by it, which may be because one does not even notice it, or because one is *intentionally* attending to it." This is not persuasive. Surely the fact that one misidentifies an object does not establish that something was not consciously perceived. The "identity" of any visual object is not settled univocally. The love that Oedipus had for Jocasta was not filial. The blue flower that is brighter in the moonlight than the yellow one that is brighter at noon offers no evidence against transitive perceptual consciousness; nor would the testimony of the honeybee whose peak spectral sensitivity is in the ultraviolet range of the spectrum. Furthermore, a sure sign that transitive perceptual consciousness is at work is the *intentional* segregation of items in visual space. As for one perceiving X "without being conscious of it," I fear some sort of theoretical special pleading is required before agreement is to be widespread.

These are fairly minor scruples, especially when weighed against those set forth by John Searle. He aims to put consciousness back in the brain, which is where Galen had it and Hippocratics long earlier. Searle improves on their conclusions with the benefit of scientific progress and by way of *states* as distinct from *places*. It is assumed there is less peculiarity in contending that consciousness is a brain *state* than that it is "in" the brain.

Talk of states and such kindred terms as processes and mechanisms has been adopted with such frequency that the terms now have a nearly protected status. But it is less than clear that they serve any purpose other than the illicit importation of conclusions into an argument that has yet to be made. The practice is habit-forming. Searle allows himself the postulation of "states" and, in no time, adds to these something called "the qualitative character of drinking beer," which, we learn, "is different from the qualitative character of listening to Beethoven's Ninth Symphony." I am reasonably confident that I know the difference between drinking beer and listening to Beethoven. I am less than sure about the "qualitative character" of either of these activities. I've never consumed a qualitative character, though I've had my share of lager. Is this a quibble? Perhaps. But quibbles are additive in this area and may reach the philosophical equivalent of critical mass.

Searle gets to the heart of his critique with his characteristic and admirable directness:

> Much of the very best work in neuroscience is an effort to explain how brain processes cause the visual experience and where and how it is realized in the brain. Astoundingly, Bennett and Hacker deny the existence of the visual experience in this sense, in the sense of quale.

Before considering the astounding denial by Bennett and Hacker, it is important to examine what I would take to be the far more astounding claim that the best work in neuroscience promises to explain how brain processes *cause* visual experience (realized somewhere in the brain). John Searle knows so well that this entire matter of causation is central to the issue at hand. He knows, therefore, that taking such causation to be settled—with good research counted on to show how it all works—will not pass philosophical muster. There is no widespread agreement as to the causal *relata* themselves (are they facts, material objects, conceptual terms, events, conditions?) or even if they must invariably exist. After all, the cause of Jane's survival was her not drinking the poison. Here the "cause" of survival is a nonevent. To get right to the main point, let us grant that if, indeed, all that marks out the domain of the mental is causally brought about by some set of "states" in the brain, then, as the maxim goes, *physics is complete* and we can begin retooling philosophers for a second career.

As a long-time resident of planet earth, I have little doubt but that the healthy and functional organization of the body, especially including the nervous system, constitutes the necessary *conditions* for what we are pleased to call our mental life, at least in its sublunary incarnations. Nonetheless, the suggestion that excitable tissue *causes* all this would be nothing short of breathtaking in an age that had not already converted science into a species of rhetoric. At just the time when leading thought in physics displays discipline and wariness in the matter of causality, here come the cognitive neuroscientists and their philosophical *famulus* wondering how anyone could hesitate to accept so obvious a mission: "how brain processes cause the visual experience and where and how it is realized in the brain"! Even granting that, in some metaphysically acceptable sense, we might claim to have established how gravitational forces *cause* the key to the front door to fall toward the center of the earth, the account works (if it does at all) insofar as earth and house keys both have mass and the magnitude of their separation can be specified in miles, inches, feet, or (reluctantly) meters. The metaphysical bar is set much higher, however, when the causal connection is between metabolic activity *anywhere* and hearing "An die Freude," let alone scoring it.

Searle expresses grave doubts about the reach of the language game. The conceptual resources widely applied by Bennett and Hacker are indebted to Wittgenstein, and Searle raises what is in fact a well-known reservation about, in his words, the Wittgensteinian mistake of confusing

> the criterial basis for the application of the mental concepts with the mental states themselves. That is, they confuse the behavioral criteria for the *ascription* of psychological predicates with the *facts ascribed by these* psychological predicates, and that is a very deep mistake.

This is far too vexing an issue to address briefly. It is uncontroversial that first-person and third-person accounts of pain are drawn from different sources. It is uncontroversial that the basis on which Smith *feels* pain in his tooth is different from the basis on which Jones judges Smith to be in pain. What approaches controversy, however, are claims to the effect that—possessed of no more than excessive firing patterns in the relevant fibers of the maxillary branch of the trigeminal nerve—Smith could be

said to have "pain" as would socially and linguistically accultur-
ated persons. I have my own doubts here, but, again, space prevents a fuller exposition. Searle, however, having noted the debts
to Wittgenstein, takes a passage from Bennett and Hacker only
to interpret it oddly. Here is the passage:

> *The criterial grounds of the ascription of a psychological predicate*
> *are . . . partly constitutive of the* meaning *of that predicate. . . .*
> *The brain does not satisfy the criteria for being a possible subject of*
> *psychological predicates.*   (p. 83)

Searle understands this as denying consciousness to brains in
virtue of the fact that brains cannot "behave." He says that Bennett
and Hacker's main point is that "brains are incapable of exhibiting the appropriate behavior." Well, yes, Bennett and Hacker do
say this. However, Searle misses the subtle (perhaps too subtle)
argument from which such conclusions are drawn. It is not that
consciousness cannot be ascribed to brains because brains are
incapable of exhibiting the appropriate behavior. Rather, the
ascriptions in question, if they are to be meaningful in the root-
sense of meaning, face the same criterial requirements faced by
any predicate. Statements to the effect that Smith is tall, brains are
wet, and Harriet is young are intelligible to the extent that "wet,"
"young," and "tall" are not drawn from the box labeled BEETLE
and visible only to the one holding it. The sense in which Smith
as an isolate could attach no meaning to his being called tall is the
sense in which "pain," too, would be improperly ascribed *even to
himself.* Unless I, too, misread Bennett, Hacker, and Wittgenstein,
the conclusion is not that brains cannot be conscious but that
utterances to that effect are as incomprehensible as claims to the
effect that brains are social democrats.

In several places, but chiefly toward the end of his interesting essay, Searle alludes to the contributions experimental science
might make to philosophical problems. Noting that questions of
"the good life," etc., are not likely to receive such benefits, he

nonetheless expects some philosophical problems to yield to scientific findings. Regrettably, the specific example he offers leaves at least one reader hopelessly perplexed. Searle's words are these:

> I do not make a sharp distinction between scientific and philosophical questions. Let me give one example to explain how my philosophical work can be helped by scientific discoveries. When I raise my arm, my conscious intention-in-action causes a physical movement of my body. But the movement also has a level of description where it is caused by a sequence of neuron firings and the secretion of acetylcholine at the axon end plates of the motor neurons. On the basis of these facts I can do a philosophical analysis to show that one and the same event must be both a qualitative, subjective, conscious event and also have a lot of chemical and electrical properties. But there the philosophical analysis ends. I need now to know how exactly it works in the plumbing.

On reading this, one must wonder what sort of "philosophical analysis" results in the conclusion that the movement of the arm has "a lot of chemical and electrical properties." At one level, it is obvious (and would have been so to preliterate cave dwellers) that arms have weight, that something under the skin of the arm increases in tension and that, when penetrated by sharp objects, these same arms exude hot red liquid. There is no question but that it is the business of experimental science to work out the details of all the subcutaneous events associated with raising one's arm. It is doubtful that a philosophical analysis would be anything but a distraction as the research team presses on with this important work. Meanwhile, the philosophically inclined might wonder just what the difference is between someone having an arm passively raised by another and someone *intentionally* achieving the same result. Without benefit of any sort of scientific research, it would be plausible to conclude that a difference of this sort will be expressed somewhere under the skin, even while recognizing that the physicochemical differences will not "explain" intentions. But when we acknowledge the careful, sustained, and informing thought that philosophy has devoted to the vexing question of volitional activity—all this long before anyone knew there were neurons or end plate potentials—the proposition that such contributors ever needed to know, "how exactly it works in the plumbing," becomes less than credible.

Searle is prepared to grant the conclusions Wittgenstein reached in the matter of the "language game" but finds them irrelevant to the project of cognitive neuroscience. Thus,

> when we are investigating the ontology of pain—not the conditions for playing the language game, but the very ontology of the phenomenon itself—we can forget about the external behavior and just find out how the brain causes the internal sensations.

I repeat my dubiety as regards "how the brain causes . . . sensations" and turn to Searle's sense of the ontology of pain, by which he means the "real" sensation itself. Of course, the living brain is never silent, and therefore the number of possible neural-phenomenal correlates is effectively limitless. The so-called classical pain pathways terminate in the thalamus, there being no cortical pain "center" as such. More narrowly, then, the question would be how thalamic nuclei "cause" pain. The nuclei are constellations of integrated cell bodies functioning as a unit. So now we refine the quest still further: How graded potentials arising from cell bodies within the thalamus cause pain.

Let us offer as an example the stretching or twisting of one's arm to the point of felt pain. The pain, of course, is felt in the arm, not the thalamus, for nothing in the brain "feels" anything. We know Jack is in pain, for he grimaces and says, "Ouch!" We know, too, that c-fibers are stimulated and that signals entering the dorsal surface of the spinal cord will journey toward the brain and the relevant thalamic nuclei. But much else is on that same journey, coming from the same arm. Moreover, there's no "Ouch!" until the discharge rates reach and exceed a critical value. Without any facial, postural, or vocal response on the part of Jack, we would have all this neurophysiological data but nowhere to go with it. It is, in the end, Jack who has the last word on the pain. But how do we know which of Jack's "signs" is the one for pain? Indeed, how does Jack know? I believe Searle has left the language game at least one phoneme too soon.

In discussing the mereological fallacy, Searle rejects that there is a fallacy at all and argues that even Wittgensteinian arguments are not at the expense of the notion that (in a special sense) "the brain thinks" or "the brain sees." As Searle puts it, Bennett and Hacker contend that the brain does not think and that thinking cannot occur in the brain. He insists that "they would need a separate argument to show that the brain cannot be the locus of such processes and I cannot find that argument." Rather,

> All that the Wittgensteinian argument requires is that the brain be part of a causal mechanism of a total system capable of producing the behavior. And that condition can still be satisfied even when certain psychological processes are located in the brain.

As the quibbles add up further, one begins to sense that talk about thinking as a "process" more or less requires one to look for "its" locus, and that only a distinct argument to the contrary will weaken what common sense makes evident here. Thinking as a "process" presumably would be, of course, a "brain process," at least if one must choose from the organs of the body. If, however, "thinking" is the word we apply to that large number of notions, expectations, beliefs, judgments, strategies, etc., with which the unforgiving minute of consciousness is filled—or if we apply it to one of those ideas with which one might be obsessed for long periods without interruption—I should think the burden would be on those contending that any of this is a "process," let alone one taking place in the brain. Consider that massless, spaceless entity *information*, in the strict sense of that which alters probabilities and the overall entropy in a system. This is no time to enter the thrilling world of quantum indeterminacy and superposition as we set about to rescue Schroedinger's cat, but it is useful nonetheless to recognize that our most developed science is far less committed to the view that real effects require places, masses, and observable "processes." It would seem that the ultimate status of physicalism will depend on just how mental life is best and most fully explained, but it is surely far too early take a firm position on the matter. The right place to begin is with our very terms of choice, making sure that we do not adopt modes of speech that virtually foreclose opportunities for unearthing our systematic ignorance.

Daniel Dennett, with whom John Searle has had his own disagreements, is nevertheless as critical of Bennett and Hacker as is Searle, but on different if overlapping grounds. Dennett's primary line of defense is to cast himself as one actually extending the work of, as he puts it, "St. Ludwig," by focusing on the behavior of robots, chess-playing computers, and even brains and their parts—behavior sufficiently like that of persons to allow predication in psychological terms. Citing his own earlier work, Dennett argues that it is precisely because there are two "levels of explanation" that we are summoned to the task of relating them, a task calling for philosophical analysis (*Content and Consciousness,* pp. 95–96). However, to assume that there are, indeed, two levels of explanation does not of itself establish that they are or can be related or, if related, that the expected form of the relationship will be causal. Were the relationship to prove to be that obtaining between, say, ambient temperature and the mean kinetic energy in the system, we would have an identity relation. If, however, the relationship is that between a street address and a particular family residence, surely nothing informative would be forthcoming simply by knowing that the Smiths are at 77

Chestnut Lane and the Joness at 79. There is a definite relationship between Bill's decision to attend the concert and the
direction in which his feet move once the car is parked. There
is a comparably definite relationship between activity in the ex-
trapyramidal pathways of Gouverneur Morris as he crossed out
"We the States" and inserted "We the People" in the preamble
to the Constitution of the United States. Would it not be droll,
however, to say that, in accounting for the individuation of rights
enjoyed by the first citizens of the United States, there are two
levels of explanation to be considered and that one of them
pertains to the extrapyramical pathways of Gouverneur Morris?
I belabor the point.

What of Kasparov and the alleged "chess-playing" computer?
At one point, the frustrated opponent of Deep Blue declared
that his adversary was simply not *playing* chess. It lacked the passion, was immune to the pressures, faced no adversary. Recall
Schiller's *Letters Upon the Aesthetic Education of Man,* where we
are told that man is never so authentically himself as when at
play. Consider the broad, various, cultural, and dispositional fac-
tors that need to be recruited in order to qualify an activity as
"play," and then array these against whatever "process" gets Deep
Blue to have the Bishop move to QP3. Deep Blue only "plays"
chess in the sense that the microwave "cooks" soup, though the
programming is vastly more complicated.

Might we be tempted to say that, if this is the correct characterization, then Kasparov, too, plays chess in the sense that the
microwave cooks soup, though the programming is vastly more
complicated? This, after all, is what makes the strong AI thesis so
interesting. Taking Dennett's "intentional stance," it is not only
permissible but of conceptual advantage to grant to Deep Blue
whatever motives, feelings, beliefs, and attitudes we think fitting
in accounting for what Kasparov is doing. In this way, Kasparov
is not "reduced" to a machine, but the machine is elevated to
the ranks of intelligent systems. If Kasparov and Deep Blue are
both concealed behind a screen, and if the relevant Turing que-

ries receive the same responses from both, then both are "intelligent" over the range of successfully answered questions. But then we enter Searle's Chinese Room and begin to think that Deep Blue is simply a card-sorting device whose "responses" are not responses at all. They are merely "outputs." The debate goes on, but only because of a form of intellectual hysteria that renders highly educated persons blind to the silliness of the original proposition: viz., that Deep Blue is *playing chess*.

Dennett draws attention to Hacker's preference for "sense" and "nonsense" as the relevant features of a philosophical argument, contrasting this with the "true" and the "false" of science. Taken with perhaps too much sobriety, Hacker's classifications here are suits too tight for day wear. Whatever it is that is fashioned in the busy kilns of science, the very fact of later corrections, revisions, and refinements makes clear that it was not "truth" to begin with; nor was it nonsense, or at least utter nonsense, except in a few glaring instances. (Heated objects do not rise owing to their taking on the substance of *levity*, and dry-eye is surely not a reliable sign of witchery). Nonsense, too, is too strong a condemnation of philosophical positions that are obtuse, overreaching, pointlessly autobiographical, or firmly humorless. (In my collegiate days, I thought that Hume's attempt to reduce the concept of causation to objects constantly conjoined in experience was a bit of Celtic wit. Only later was I forced to accept the distressing conclusion that he meant it!) But we do not mount the right sort of criticism against Hacker's classifications merely by taking exception to his particular choice of words. The savants of the Vienna circle, dutiful in their celebratory Verein Ernst Mach, were inclined to regard all nonempirical claims as literal nonsense. Those witty Scriblerians—Pope, Swift, Arbuthnot—read Locke's pages on personal identity and concluded that he had gone off the philosophical deep end. Let's agree that philosophical treatises on significant topics must strive for clarity, accessibility, coherence, and, alas, an evident respect for the chosen topics. Those that fail in these respects are an offense to common sense.

It's not that common sense is the ultimate arbiter; only that it is the arbiter that must finally be won over if the treatise is to have influence beyond the seminar room. If Spassky and Kasparov are doubtful as to whether computers are "playing" chess, is it not Dennett who must rethink the matter?

In this same connection, Dennett claims to unmask Ryle and Wittgenstein and show them (and Hacker) to be less than sincere in suggesting that there are "rules" that govern usage in philosophical, let alone in ordinary discourse. Ryle's *category mistake* and the promise of a "logic of existence" he treats as a bluff. For all the labor devoted to syntax, the linguists still fumble with "The cat climbed down the tree." So be it. Nor can we rule out "Breakfast was a delight to the hypothalamus, for witness how its electrical behaviour was sated as the meal progressed." What is broken here is not a law but a convention—which is to say a *rule* that, unlike a law, results not in being arraigned but in being misunderstood. When such expressions become habitual, the misunderstandings become systematic, bloated with unintended paradox, rife with unintended implications, occasionally relieved by unintended humor.

Of course the "conventions" within the Politburo are different from those within the British Parliament. Where meaning itself must be nailed down, it matters just whose conventions are to prevail. Thus arises the dilemma of deciding just how much obeisance is to be paid to Folk Psychology. Dennett warns neuroscientists to exercise "the utmost caution" when trafficking in the terms of this psychology, for, as he says, the "presuppositions of use can subvert their purposes." How? By turning "otherwise promising empirical theories and models into thinly disguised nonsense." If I understand what is meant here by the "presuppositions of use," I would hazard the opinion that the core presupposition is that what Bennett and Hacker call "ordinary psychological description" must make possible the entire range of practical and meaningful interactions among native speakers. Clearly, if their unschooled patois is but thinly disguised nonsense, it is doubtful that they would have the linguistic resources even to benefit from philosophical enlightenment. Perhaps some day the rider on the Clapham Omnibus will agree to speak in terms of his "cortical triune system," sheepishly retreating from the old, thinly disguised nonsense of red apples, green turf, and blue skies. One wonders, however, just in case the human race had been born into such a language game, how the neuroscientists would match up anything in the "cortical triune system" with—yes—the world as actually *seen*. If, indeed, the discourse of the Folk might corrupt empirical theories and models, this must be so chiefly because the theories and models have no special relationship to this very discourse, which, after all, is the discourse of lived life. Perhaps this helps explain why the theories and models now on offer are no more than models of data, no more than efficient summaries of overly antiseptic observations smoothed by arguable statistical manipulations and presented as a highly integrated model of—*no one*, not even a brain.

Dennett defends himself against the charge of committing the mereological fallacy by citing his own earlier works in which he makes the distinction between personal and subpersonal levels of explanation. He speaks of himself as a pioneer in this regard. I might have extended the laurel to Aristotle who reminds us that, when explaining, for example, anger, one might speak of changes in the temperature of the blood or, instead, the reaction to having been slighted:

> A physicist would define an affection of the soul differently from a dialectician; the latter would define e.g., anger as the appetite for returning pain for pain ... while the former would define it as a boiling of the blood or warm substance surrounding the heart.[2]

Actually, we might credit Aristotle with the earliest cautions against the fallacy for, in the same treatise, he says,

To say that it is the soul which is angry is as inexact as it would be to say that it is the soul that weaves webs or builds houses. It is doubtless better . . . to say that it is the man who does this with his soul.[3]

The larger distinction between explanations grounded in causes and those grounded in reasons is venerable of course and no less controversial for being widely recognized as philosophically consequential. The mereological fallacy, however, manifests itself in different ways. It is transparent when one suggests that the birthday cake was cut by efferent nerves exiting from cervical levels 5–8 of the spinal cord. It is at work also, however, when treating Ronald's reason for acting as the sum of a large number of mini-reasons. Thus, the right explanation of Ronald buying the Prius is that he seeks to gain mechanical advantage greater than that achieved by walking. That Dennett is a victim of this is established with commendable clarity in his own words:

> We don't attribute *fully fledged* belief . . . to the brain parts—that would be a fallacy. No. We attribute an attenuated sort of belief . . . to these parts.

The illustration offered is that of the child who "sort of" believes that daddy is a doctor. This is unconvincing. There may be some hesitation in the matter of belief, but there isn't a "part" of a belief. However, the example itself must finally yield parts of beliefs, for to "attenuate" a belief through the action of a physical system (e.g., parts of brains) is to change its value along some sort of physical continuum, and this is finally to work on its "parts." If there is something of the Red Queen in talk about the brain holding beliefs, she reappears with a vengeance when part of a brain must have attenuated beliefs.

Dennett is especially forceful in attempting to rebut arguments to the effect that the brain forms images of one sort or another. He correctly declares that whether or not structures within the brain are organized in such a way as to *function* as image makers of a sort is an empirical question beyond the ambit of philosophical modes of analysis. Just how the external world is reacted to by constellations of interconnected neurons is a core question in the brain sciences and has been especially fruitfully engaged through research on the visual system. No one seriously thinks the visual world is projected onto the brain as an *image*—and surely there are no "images" of odors or sounds. Rather, one searches for isomorphic relations between the optical features of the visible world and the neuroelectric patterns associated with their inspection. All this granted, we are then left with a very different question: not how some neuroelectric algorithm treats or "codes" optical properties of the visible world, but the relationship between that coding and what the percipient claims to be (visually) the case. To think that *this* question is fit for empirical investigation is to miss the very point of the question itself, for nothing observed at the level of functional neuroanatomy "sees" in *any* sense, even an attenuated one.

Less time is called for here in considering Dennett's defense of LeDoux and others in the cognitive neuroscience *collegium*. To say that the "brain" may know of a danger before it or we know what it is can be no more than a corruption of language and, at the level of scientific explanation, a woeful blunting of Occam's razor. The neonatal macaque has cells in the auditory cortex that respond to the distress cries of that species. They do not "know" anything anymore than a resistance-capacitance circuit "knows" that a fridge has been installed, whereupon it responds with a voltage drop! Much prewiring and some hard-wiring goes into the formation of creatures facing a perilous world. They are fitted out by nature to do by instinct or reflexively what cannot wait till advanced degrees are earned. This is the equipment that bypasses *all* learning and, therefore, all knowledge. Dennett insists on fitting any number and variety of facts into a conceptual container too elastic to have shape and too thin for the weight of the real problems. His defense is to point to all the little facts that can be crammed into the thing, but it is generally regarded as rhetorically ineffective merely to repeat sentences that aroused criticism in the first place. He and many others in the cognitive neuroscience *movement* (for it has all the features of a movement) have adopted an idiom strange to ears beyond their own, but derive reassurance from the appearance of just these phrases in all of their books and articles. Dennett's validating his choice of odd locutions by citing the frequency with which he uses them is—to bend an example from Wittgenstein—"as if someone were to buy several copies of the morning newspaper to assure himself that what it said was true."[4]

What of Bennett and Hacker and Hacker especially? I reviewed their book most favorably in *Philosophy*, and nothing in the replies of John Searle or Daniel Dennett would cause me to reconsider my earlier judgment. I regarded the aims of the authors to be precisely those that constitute the very mission of philosophy, which, in its largest projection, is nothing less than the criticism of life and, in its more modest ambitions, a critical inquiry into our core epistemic claims. History makes all too clear the consequences of trading this mission in for a loftier position in that hierarchy within which science itself is located. There is a quite significant respect in which Sophocles, through the instrument of *Antigone*, defends the moral foundations of all law against the pretensions of a king. But Sophocles did not accomplish the work to which Aristotle, Cicero, Aquinas, and others in the Natural Law tradition applied themselves. The Hippocratics wisely collapsed the very notion of a divine malady, first by courteously acknowledging that the gods bring about everything and then treating each disease as no more or less divine than any other. Without recommending conceptual straightjackets, I would endorse a regimen of restraint and focus as serious persons (whether philosophers, scientists, or just interested parties) consider the reach and the authority of philosophical and scientific reflections on the *Lebensweld*. One can and should admire and be instructed by *Antigone*, without then requiring its inclusion in legal briefs to settle disputes in maritime law. One can and should admire the insulation the Hippocratics strived to create for medicine against ritual religious convictions without insisting that prayer be forbidden in the emergency room. And one can and should admire the detailed, repeatable, and exacting research of the neuroscience community without developing a sceptical attitude toward common sense and toward the otherwise insistent recognition of ourselves as—ourselves.

Whether or not Peter Hacker's philosophical analysis of the conceptual blunders of today's neuroscientists is entirely sound or doomed by later reckonings, there is no question but that it is at once faithful to the philosophical mission and in full command of the resources that philosophy has fashioned for that mission. He does not seek honorary membership in the Society of Neuroscience, nor does he pretend to enrich that already impressive database on which scientific progress must be built. In point of fact, the really significant discoveries in cognitive neuroscience have been made by a small legion of specialists largely unknown to readers of the *TLS* and the *New York Review of Books*. No grand "neurophilosophy" was served up by those who put vision science on the map of truly first-rate science. The names will mean nothing to nearly all readers of my humble prose: Selig Hecht, M.H. Pirenne, Clarence Graham, H.K. Hartline, George Wald. Pitts and McCulloch stayed close to their mathematics and conjured circuits that achieve remarkable outcomes when cleverly designed. Pavlov was productive when laboring over the chemistry of digestion but became something of a hack when attempting to translate all of psychology into the language of "cerebrodynamics." DuBois was wiser when facing this *quaestio vexata* and concluding, with joy in his voice, I'm sure, *IGNOR-ABIMUS!*

If Hacker seeks membership in any circle of orthodoxies, where fees are exacted in the form of clichés, it would be that of anti-*Cartesianism*. By cliché I mean no more than a hackneyed expression or maxim. "God is good," is such a cliché, and faithful persons will take the frequency of the assertion as a measure of its truth. "I reject Cartesianism" is another hackneyed claim and it, too, might be recording an insight into some deep or higher truth. But Cartesianism means different things to different members of the anti-Cartesian circle. In some untidy way, it is generally blamed for a two-substance ontology combined with a "theater-of-the-mind" theory. Both dualism and the inner theater are then mirthfully dismissed as evidence of philosophical innocence.

I assume it is still permissible to remind all that Descartes was the founder of analytic geometry—a veritable master of the science of optics—one who fully comprehended the science of his day and who, in his correspondence and letters, engaged the best minds of an age that was populated by great minds. Little criticism against his views has ever been advanced in the most recent two centuries that had not been anticipated by Hobbes, Gassendi, and Father Mersenne, with whom Descartes carried on a spirited debate in print. As he made clear to Princess Elizabeth, it was useful for him to adopt in his writing phrases and analogies that were perhaps overly philosophical in order not to be misunderstood. Knowing what he did about matter, he was satisfied that the essential character of rational and perceptual life could not be derived from matter in any combination. Is there something really ridiculous about a dualistic ontology that contrasts extended and unextended entities? I don't think so; in point of fact, *no one thinks so*, for thinking itself rules it out.

Is this an argument for dualism? On and off over a period of perhaps fifty years I've pondered just how many kinds of different sorts of "stuff" might be constitutive of all reality. I've been able to comprehend two kinds, which, for want of better words, I would call physical sorts of stuff and whatever it is that grounds the moral, aesthetic, rational, and emotional dimensions of my life. Oh, call it "mental." As it happens, there's not much of the electromagnetic spectrum that is visible to me; only the "stuff" that has wavelengths of some 3,600 to about 7,600 angstroms. Just in case my capacity to comprehend the entirety of reality is limited in a manner akin to that which restricts my vision, it's best to leave un-answered questions about just how many distinguishable types of entities comprise "all of what there is." I don't know the number. Dennett doesn't know. Hacker doesn't know. But what if "physics is complete"? Doesn't that settle it? As a previous U.S. president might say, "It all depends on what you mean by 'complete.'"

Hacker is in the Wittgensteinian school of philosophy whose major defenders and critics are drawn from the ranks of well-schooled philosophers. Whether or not one regards philosophical issues as "puzzles" to be worked out or "problems" to be solved is a large topic. On either construction, clarity and consistency of expression are essential. No one would seriously declare that, not being Wittgensteinian, there is no obligation to analyse the cultural and linguistic apparatus by which concepts are created and shared.

Hacker writes with precision, so much so as to lean toward that etymological cousin, *preciousness*. He is careful. Readers might react to such care the way we react to drivers who never exceed the posted speed. When he says that

> conceptual truths delineate the logical space within which facts are located. They determine what makes sense.

he might be judged as undervaluing facts or giving philosophy ruling power over them. He is doing no such thing. The cosmos is ablaze with facts, the great plurality of them beyond our senses and even our ken. Out of that fierce and brilliant fire, we pull a few bits—the visible or nearly visible ones—and begin to weave a story. On rare occasions, the story is so systematic, so true to the bits in hand, that other stories flow from the first, and then others, and soon we are possessed of utterly prophetic powers as to which ones will come out next. It is the philosopher, how-ever, who must put the brakes on the enthusiasms of the story-tellers, for, left to their own devices, they might conjure a future that vindicates only our current confusions.

NOTES

THE INTRODUCTION TO
Philosophical Foundations of Neuroscience

The following is the unaltered text of the preface of *Philosophical Foundations of Neuroscience*, save for cutting the last two paragraphs and the elimination of cross-references, which have been replaced, where necessary, by notes. (Subsequent references to the book are flagged PFN.)

1. Methodological objections to these distinctions are examined in the sequel and, in further detail, in PFN, chapter 14.
2. Chapter 1 of PFN accordingly begins with a historical survey of the early development of neuroscience.
3. Chapter 2 of PFN is accordingly dedicated to a critical scrutiny of their conceptual commitments.
4. PFN §3.10.
5. See below, in the excerpt from chapter 3 of PFN. The original chapter is much longer than the excerpt here supplied and the argument correspondingly more elaborate.
6. Reductionism is discussed in PFN chapter 13.
7. In PFN chapter 14.
8. See PFN §14.3.
9. See PFN chapters 1 and 2.
10. Examples that arguably render the research futile are scrutinized in PFN §6.31, which examines mental imagery, and PFN §8.2, which investigates voluntary movement.
11. Examples are given in the discussions of memory in PFN §§5.21–5.22 and of emotions and appetites in PFN §7.1.
12. We address methodological qualms in detail in PFN chapter 3, § 3 (this volume) and in PFN chapter 14.

AN EXCERPT FROM CHAPTER 3

These pages consist of the unaltered text of PFN, pp. 68–80, save for cross-references, which have been relegated to notes where necessary.

1. F. Crick, *The Astonishing Hypothesis* (Touchstone Books, London, 1995), pp. 30, 32f., 57.

2. G. Edelman, *Bright Air, Brilliant Fire* (Penguin Books, London, 1994), pp. 109f., 130.

3. C. Blakemore, *Mechanics of the Mind* (Cambridge University Press, Cambridge, 1977), p. 91.

4. J.Z. Young, *Programs of the Brain* (Oxford University Press, Oxford, 1978), p. 119.

5. A. Damasio, *Descartes' Error—Emotion, Reason and the Human Brain* (Papermac, London, 1996), p. 173.

6. B. Libet, 'Unconscious cerebral initiative and the role of conscious will in voluntary action', *The Behavioural and Brain Sciences* (1985) **8**, p. 536.

7. J.P. Frisby, *Seeing: Illusion, Brain and Mind* (Oxford University Press, Oxford, 1980), pp. 8f. It is striking here that the misleading philosophical idiom associated with the Cartesian and empiricist traditions, namely talk of the 'outside' world, has been transferred from the mind to the brain. It was misleading because it purported to contrast an inside 'world of consciousness' with an outside 'world of matter'. But this is confused. The mind is not a kind of place, and what is idiomatically said to be *in* the mind is not thereby spatially located (cp. 'in the story'). Hence too, the world (which is not 'mere matter', but also living beings) is not *spatially* 'outside' the mind. The contrast between what is in the brain and what is outside the brain is, of course, perfectly literal and unobjectionable. What is objectionable is the claim that there are 'symbolic descriptions' in the brain.

8. R.L. Gregory, 'The Confounded Eye', in R.L. Gregory and E.H. Gombrich eds. *Illusion in Nature and Art* (Duckworth, London, 1973), p. 50.

9. D. Marr, *Vision, a Computational Investigation into the Human Representation and Processing of Visual Information* (Freeman, San Francisco, 1980), p. 3 (our italics).

10. P.N. Johnson-Laird, 'How could consciousness arise from the computations of the brain?' in C. Blakemore and S. Greenfield eds. *Mindwaves* (Blackwell, Oxford, 1987), p. 257.

11. Susan Greenfield, explaining to her television audiences the achievements of positron emission tomography, announces with wonder that for the first time it is possible *to see thoughts*. Semir Zeki informs the Fellows of the Royal Society that the new millennium belongs to neurobiology, which will, among other things solve the age old problems of philosophy (see S. Zeki, 'Splendours and miseries of the brain', *Phil. Trans. R. Soc. Lond.* B (1999), **354**, 2054). See PFN §14.42.

12. L. Wittgenstein, *Philosophical Investigations* (Blackwell, Oxford, 1953), §281 (see also §§282–4, 357–61). The thought fundamental to this remark was developed by A.J.P. Kenny, 'The Homunculus Fallacy' (1971), repr. in his *The Legacy of Wittgenstein* (Blackwell, Oxford, 1984), pp. 125–36. For the detailed interpretation of Wittgenstein's observation, see P.M.S. Hacker, *Wittgenstein: Meaning and Mind, Volume 3 of an Analytical Commentary on the Philosophical Investigations* (Blackwell, Oxford, 1990), Exegesis §§281–4, 357–61 and the essay entitled 'Men, Minds and Machines', which explores some of the ramifications of Wittgenstein's insight. As is evident from [PFN] Chapter 1, he was anticipated in this by Aristotle (DA 408^{b}2–15).

13. Kenny (ibid., p. 125) uses the term 'homunculus fallacy' to signify the conceptual mistake in question. Though picturesque, it may, as he admits, be misleading, since the mistake is *not* simply that of ascribing psychological predicates to an imaginary homunculus in the head. In our view, the term 'mereological fallacy' is more apt. It should, however, be noted that the error in question is not merely the fallacy of ascribing to a part predicates that apply only to a whole, but is a special case of this more general confusion. As Kenny points out, the misapplication of a predicate is, strictly speaking, not a fallacy, since it is not a form of invalid reasoning, but it leads to fallacies (ibid., pp. 135f.). To be sure, this mereological confusion is common among psychologists as well as neuroscientists.

14. Comparable mereological principles apply to inanimate objects and some of their properties. From the fact that a car is fast it does not follow that its carburettor is fast, and from the fact that a clock

tells the time accurately, it does not follow that its great wheel tells the time accurately.

15. But note that when my hand hurts, I am in pain, not my hand. And when you hurt my hand, you hurt me. Verbs of sensation (unlike verbs of perception) apply to parts of the body, i.e. our body is sensitive and its parts may hurt, itch, throb, etc. But the corresponding verb phrases incorporating nominals, e.g. 'have a pain (an itch, a throbbing sensation)' are predicable only of the person, not of his parts (in which the sensation is located).

16. See Simon Ullman, 'Tacit Assumptions in the Computational Study of Vision', in A. Gorea ed. *Representations of Vision, Trends and Tacit Assumptions in Vision Research* (Cambridge University Press, Cambridge, 1991), pp. 314f. for this move. He limits his discussion to the use (or, in our view, misuse) of such terms as 'representation' and 'symbolic representation'.

17. The phrase is Richard Gregory's, see 'The Confounded Eye' in R.L. Gregory and E.H. Gombrich eds. *Illusion in Nature and Art* (Duckworth, London, 1973), p. 51.

18. See C. Blakemore, 'Understanding Images in the Brain', in H. Barlow, C. Blakemore and M. Weston-Smith eds. *Images and Understanding* (Cambridge University Press, Cambridge, 1990), pp. 257–83.

19. S. Zeki, 'Abstraction and Idealism', *Nature* 404 (April 2000), p. 547.

20. J.Z. Young, *Programs of the Brain* (Oxford University Press, Oxford, 1978), p. 192.

21. Brenda Milner, Larry Squire and Eric Kandel, 'Cognitive Neuroscience and the Study of Memory', *Neuron* 20 (1998), p. 450.

22. For detailed discussion of this questionable claim, see PFN §5.22.

23. Ullman, ibid., pp. 314f.

24. Marr, ibid., p. 20.

25. Marr, ibid., p. 21.

26. Marr, ibid.

27. For further criticisms of Marr's computational account of vision, see PFN §4.24.

28. Frisby, ibid., p. 8.

29. Roger Sperry, 'Lateral Specialization in the Surgically Separated Hemispheres', in F.O. Schmitt and F.G. Worden eds. *The Neurosciences Third Study Programme* (MIT Press, Cambridge, Mass., 1974), p.

11 (our italics). For detailed examination of these forms of description, see PFN §14.3.

30. Blakemore, 'Understanding Images in the Brain', p. 265. It should be noted that what is needed in order to recognize the order in the brain is *not* a set of *rules*, but merely a set of regular correlations. A rule, unlike a mere regularity, is a standard of conduct, a norm of correctness against which behaviour can be judged to be right or wrong, correct or incorrect.

31. J.Z. Young, *Programs of the Brain* (Oxford University Press, Oxford, 1978), p. 52.

32. Blakemore, ibid., pp. 265–7.

33. J.Z. Young, *Programs of the Brain*, p. 11.

34. Just how confusing the failure to distinguish a rule from a regularity, and the normative from the causal, is evident in Blakemore's comments on the Penfield and Rasmussen diagram of the motor 'homunculus'. Blakemore remarks on 'the way in which the jaws and hands are vastly over-represented' ('Understanding Images in the Brain', p. 266, in the long explanatory note to Fig. 17.6); but that would make sense only if we were talking of a map with a misleading method of projection (in this sense we speak of the relative distortions of the Mercator (cylindrical) projection. But since all the cartoon drawing represents is the relative number of cells causally responsible for certain functions, *nothing* is, or could be, 'over-represented'. For, to be sure, Blakemore does not mean that there are more cells in the brain causally correlated with the jaws and the hands than there ought to be!

AN EXCERPT FROM CHAPTER 10

1. Ned Block, 'Qualia', in S. Guttenplan ed. *Blackwell Companion to the Philosophy of Mind* (Blackwell, Oxford, 1994), p. 514.

2. .R. Searle, 'Consciousness', *Annual Review*, p. 560.

3. Searle, ibid., p. 561.

4. Chalmers, *The Conscious Mind* (Oxford University Press, Oxford, 1996), p. 4.

5. D.J. Chalmers, *The Conscious Mind*, p. 10.

6. I. Glynn, *An Anatomy of Thought*, p. 392.

7. A. Damasio, *The Feeling of What Happens*, p. 9. Note that there is here an unargued assumption that colour and sound are not properties of objects but of sense-impressions.

8. G. Edelman and G. Tononi, *Consciousnes —How Matter Becomes Imagination*, p. 157.

9. E. Lomand, 'Consciousness', in *Routledge Encyclopaedia of Philosophy* (Routledge, London, 1998), vol. 2, p. 581.

10. Searle, *The Mystery of Consciousness*, p. xiv.

11. T. Nagel, 'What is it like to be a bat?', repr. in *Mortal Questions* (Cambridge University Press, Cambridge, 1979), p. 166.

12. Nagel, ibid., p. 170n.

13. Nagel, ibid., p. 170.

14. M. Davies and G.W. Humphreys ed. *Consciousness* (Blackwell, Oxford, 1993), p. 9.

15. Edelman and Tononi, *Consciousness— How Matter becomes Imagination*, p. 11.

16. Chalmers, *The Conscious Mind*, p. 4.

17. Cf. Searle, *The Mysteries of Consciousness*, p. 201.

AN EXCERPT FROM CHAPTER 14

1. See, for example, the discussion of voluntary movements in PFN §8.2.

2. See, for example, the discussion of mental imagery in PFN §6.31.

3. See, for example, PFN §14.3.

4. Discussed in PFN §4.23.

5. As argued in PFN §§6.3–6.31.

6. See PFN §2.3.

NEUROSCIENCE AND PHILOSOPHY

1. There are important criticisms of the use in this way of terms such as "storage" and "memory"; see Bennett and Hacker, *Philosophical Foundations of Neuroscience* (Oxford: Blackwell, 2003), pp. 158–71.

2. Professor Dennett suggests in his note 15 that at the APA meeting Bennett expressed "utter dismay with the attention-getting hypotheses and models of today's cognitive neuroscientists and made it clear that he thought it was all incomprehensible. With an informant like Bennett, it is no wonder that Hacker was unable to find anything of value in cognitive neuroscience." He also suggests that I am clearly caught up in that "mutual disrespect" that occurs between synaptic neuroscientists and cognitive neuroscientists. This is not correct. First, David Marr is held up to be a cognitive neuroscientist of genius in textbooks on the subject (see Gazzaniga, Ivry, and Mangun 2002:597); I have published papers on synaptic network theory in the spirit of Marr's work and do not see this in any way as showing illogical hostility to the cognitive neurosciences (see, for example, Bennett, Gibson, and Robinson 1994). A forthcoming book by Hacker and myself, *History of Cognitive Neuroscience*, would not have been written if we had been caught up in irrational hostility to cognitive neuroscience. Second, I did not claim at the APA meeting that the "models of today's cognitive neuroscientists" are "all incomprehensible." Rather I stressed the extreme complexity of the biology being modeled and the resultant paucity of our biological knowledge. This makes it very difficult to build models that illuminate synaptic network functions. The examples offered to support this view are given in the second section of this chapter. However, I did go on to say that it seems strange that such networks and collections of networks should be said to "see," "remember," etc., that is, possess the psychological attributes of human beings (see the third section).

PHILOSOPHY AS NAIVE ANTHROPOLOGY

1. My purpose in *Content and Consciousness,* in 1969, was "to set out the conceptual background against which the whole story must

be told, to determine the constraints within which any satisfactory theory must evolve (p. ix) ... [to develop] the notion of a distinct mode of discourse, the language of the mind, which we ordinarily use to describe and explain our mental experiences, and which can be related only indirectly to the mode of discourse in which science is formulated" (p. x).

2. Although earlier theorists—e.g., Freud—spoke of *folk psychology* with a somewhat different meaning, I believe I was the first, in "Three Kinds of Intentional Psychology" (1978), to propose its use as the name for what Hacker and Bennett call "ordinary psychological description." They insist that this is not a *theory*, as do I.

3. See my discussion of this in "A Cure for the Common Code," in *Brainstorms (1978)* and, more recently, in "Intentional Laws and Computational Psychology" (section 5 of "Back from the Drawing Board") in Dahlbom, ed., *Dennett and His Critics*, 1993.

4. The list is long. See, in addition to the work cited in the earlier footnotes, my critiques of work on imagery, qualia, introspection, and pain in *Brainstorms*. I am not the only theorist whose work anticipatory to their own is overlooked by them. For instance, in their discussion of mental imagery, they reinvent a variety of Zenon Pylyshyn's points without realizing it. Bennett and Hacker are not the first conceptual analysts to frequent these waters, and most, if not quite all, of their points have been aired before and duly considered in literature they do not cite. I found nothing new in their book.

5. Their appendix devoted to attacking my views is one long sneer, a collection of silly misreadings, ending with the following: "If our arguments hold, then Dennett's theories of intentionality and of consciousness make no contribution to the philosophical clarification of intentionality or of consciousness. Nor do they provide guidelines for neuroscientific research or neuroscientific understanding" (p. 435). But there are no arguments, only declarations of "incoherence." At the APA meeting during which this essay was presented, Hacker responded with more of the same. It used to be, in the Oxford of the sixties, that a delicate shudder of incomprehension stood in for an argument. Those days have passed. My advice to Hacker: If you find these issues incomprehensible, try harder. You've hardly begun your education in cognitive science.

6. Hornsby 2000. Hacker's obliviousness to my distinction cannot be

attributed to myopia; in addition to Hornsby's work, it has also been discussed at length by other Oxford philosophers: *e.g.,* Davies 2000; Hurley, *Synthese,* 2001; and Bermudez, "Nonconceptual Content: From Perceptual Experience to Subpersonal Computational States," *Mind and Language,* 1995.

7. See also "Conditions of Personhood" in *Brainstorms.*

8. See also the discussion of levels of explanation in *Consciousness Explained* (1991).

9. At the APA meeting at which this essay was presented, Searle did not get around to commenting on this matter, having a surfeit of objections to lodge against Bennett and Hacker.

10. For a philosopher who eschews truth and falsehood as the touchstone of philosophical propositions, Hacker is remarkably free with unargued bald assertions to the effect that so-and-so is mistaken, that such-and-such is wrong, and the like. These *obiter dicta* are hard to interpret without the supposition that they are intended to be true (as contrasted with false). Perhaps we are to understand that only a tiny fraction of his propositions, the specifically philosophical propositions, "antecede" truth and falsehood while the vast majority of his sentences are what they appear to be: assertions that aim at truth. And as such, presumably, they are subject to empirical confirmation and disconfirmation.

11. In *Sweet Dreams: Philosophical Obstacles to a Science of Consciousness* (2005), I describe some strains of contemporary philosophy of mind as *naive aprioristic autoanthropology* (pp. 31–35). Hacker's work strikes me as a paradigm case of this.

12. Notice that I am not saying that autoanthropology is always a foolish or bootless endeavor; I'm just saying that it is an empirical inquiry that yields results—when it is done right—about the intuitions that the investigators discover in themselves, and the implications of those intuitions. These can be useful fruits of inquiry, but it is a further matter to say under what conditions any of these implications should be taken seriously as guides to the truth on any topic. See *Sweet Dreams* for more on this.

13. *The Claim of Reason: Wittgenstein, Skepticism, Morality, and Tragedy* (1979; 2d ed. 1999).

14. Can a philosopher like Hacker be *right* even if not aiming at the truth?

15. Presumably Bennett, a distinguished neuroscientist, has played in-formant to Hacker's anthropologist, but then how could I explain Hacker's almost total insensitivity to the subtleties in the *patois* (and the models and the discoveries) of cognitive science? Has Hacker chosen the wrong informant? Perhaps Bennett's research in neuro-science has been at the level of the synapse, and people who work at that subneuronal level are approximately as far from the disciplines of cognitive science as molecular biologists are from field etholo-gists. There is not much communication between such distant en-terprises, and even under the best of circumstances there is much miscommunication—and a fair amount of mutual disrespect, sad to say. I can recall a distinguished lab director opening a workshop with the following remark: "In our lab we have a saying: if you work on one neuron, that's neuroscience; if you work on two neurons, that's psychology." He didn't mean it as a compliment. Choosing an unsympathetic informant is, of course, a recipe for anthropological disaster. (Added after the APA meeting:) Bennett confirmed this surmise in his opening remarks; after reviewing his career of research on the synapse, he expressed his utter dismay with the attention-get-ting hypotheses and models of today's cognitive neuroscientists and made it clear that he thought it was all incomprehensible. With an informant like Bennett, it is no wonder that Hacker was unable to find anything of value in cognitive neuroscience.

16. See my *Content and Consciousness*, p. 183.

17. To take just one instance, when Hacker deplores my "barbaric nomi-nal 'aboutness'" (p. 422) and insists that "opioid receptor are no more *about* opioids than cats are about dogs or ducks are about drakes" (p. 423), he is of course dead right: the elegant relation between opi-oids and opioid receptors isn't fully fledged aboutness (sorry for the barbarism), it is mere proto-aboutness (ouch!), but that's *just* the sort of property one might treasure in a mere part of some mereological sum which (properly organized) could exhibit *bona fide*, *echt*, philo-sophically sound, paradigmatic … intentionality.

18. In Hacker's narrow sense.

19. This has been an oft-recurring theme in critical work in cognitive science. Classic papers go back to William Woods's "What's in a Link?" (in Bobrow and Collins, *Representation and Understanding*,

1975) and Drew McDermott's "Artificial Intelligence Meets Natural Stupidity," in Haugeland, *Mind Design* (1981), through Ulrich Neisser's *Cognition and Reality* (1975) and Rodney Brooks's "Intelligence Without Representation," *Artificial Intelligence* (1991). They continue to this day, including contributions by philosophers who have done their homework and know what the details of the issues are.

20. Hacker and Bennett say: "It would be misleading, but otherwise innocuous, to speak of maps in the brain when what is meant is that certain features of the visual field can be mapped on to the firings of groups of cells in the 'visual' striate cortex. But then one cannot go on to say, as Young does, that the brain makes use of its maps in formulating its hypotheses about what is visible" (p. 77). But that is just what makes talking about maps perspicuous: that the brain *does* make use of them *as* maps. Otherwise, indeed, there would be no point. And that is why Kosslyn's pointing to the visible patterns of excitation on the cortex during imagery is utterly inconclusive about the nature of the processes underlying what we call, at the personal level, visual imagery. See Pylyshyn's recent target article in *BBS (April 2002)* and my commentary, *"Does Your Brain Use the Images on It, and If So, How?"*

21. "Philosophers should not find themselves having to abandon pet theories about the nature of consciousness in the face of scientific evidence. They should have no pet theories, since they should not be propounding empirical theories that are subject to empirical confirmation and disconfirmation in the first place. Their business is with concepts, not with empirical judgments; it is with the forms of thought, not with its content; it is with what is logically possible, not with what is empirically actual; with what does and does not make sense, not with what is and what is not true" (p. 404). It is this blinkered vision of the philosopher's proper business that permits Hacker to miss the mark so egregiously when he sets out to criticize the scientists.

22. For an example of such a *type* of explanation, see my simplified explanation of how Shakey the robot tells the boxes from the pyramids (a "personal level" talent in a robot) by (subpersonally) making line drawings of its retinal images and then using its line semantics

program to identify the telltale features of boxes, in *Consciousness Explained*.

23. Bennett and Hacker's "Appendix 1: Daniel Dennett" does not deserve a detailed reply, given its frequent misreadings of passages quoted out of context and its apparently willful omission of any discussion of the passages where I specifically defend against the misreadings they trot out, as already noted. I cannot resist noting, however, that they fall for the creationist canard they presume will forestall any explanations of biological features in terms of what I call the design stance: "Evolution has not *designed* anything—Darwin's achievement was to displace explanation in terms of design by evolutionary explanations" (p. 425). They apparently do not understand how evolutionary explanation works.

PUTTING CONSCIOUSNESS BACK IN THE BRAIN

I am indebted to Romelia Drager, Jennifer Hudin, and Dagmar Searle for comments on earlier drafts of this article.

1. For example, John R. Searle, *The Rediscovery of the Mind* (Cambridge: MIT Press, 1992).

2. For possible counterevidence to this claim, see Christof Koch's discussion of "the Halle Berry neuron," e.g., *New York Times,* July 5, 2005.

3. John R. Searle, *Rationality in Action* (Cambridge: MIT Press, 2001).

THE CONCEPTUAL PRESUPPOSITIONS OF COGNITIVE NEUROSCIENCE

1. M. R. Bennett and P. M. S. Hacker, *Philosophical Foundations of Neuroscience* (Oxford: Blackwell, 2003); page references to this book will be flagged PFN.

2. Professor Searle asserts that a conceptual result is significant only as a part of a general theory (p. 122). If by "a general theory" he means an overall account of a conceptual network, rather than mere piecemeal results, we agree. Our denial that our general ac-

counts are theoretical is a denial that they are logically on the same level as scientific theories. They are descriptions, not hypotheses; they are not confirmable or refutable by experiment; they are not hypothetico-deductive and their purpose is neither to predict nor to offer causal explanations; they do not involve idealizations in the sense in which the sciences do (e.g., the notion of a point mass in Newtonian mechanics) and they do not approximate to empirical facts within agreed margins of error; there is no discovery of new entities and no hypothesizing entities for explanatory purposes.

3. Professor Dennett seemed to have difficulties with this thought. In his criticisms (p. 79), he quoted selectively from our book: "Conceptual questions antecede matters of truth and falsehood . . . " (PFN 2, see p. 4, this volume) "What truth and falsity is to science, sense and nonsense is to philosophy" (PFN 6, see p. 12, this volume). From this he drew the conclusion that in our view, philosophy is not concerned with truth at all. However, he omitted the sequel to the first sentence:

They are questions concerning our forms of representation, not questions concerning the truth or falsehood *of empirical statements*. These forms are presupposed by true (and false) scientific statements, and by correct (and incorrect) scientific theories. They determine not what is empirically true or false, but rather what does and does not make sense.

(PFN 2, see p. 4, this volume; emphasis added)

He likewise omitted the observation on the facing page that neuroscience is discovering much concerning the neural foundations of human powers, "but its discoveries in no way affect *the conceptual truth* that these powers and their exercise ... are attributes of human beings, not of their parts" (PFN 3, see p. 6, this volume; emphasis added). As is patent, it is our view that philosophy is concerned with conceptual truths and that conceptual truths determine what does and does not make sense.

4. Professor Paul Churchland proposes, as a consideration against our view, that "since Quine, the bulk of the philosophical profession has been inclined to say 'no'" to the suggestion that there are "necessary truths, constitutive of meanings, that are forever beyond empirical

or factual refutation." "Cleansing Science," *Inquiry* 48 (2005): 474. We doubt whether he has done a social survey (do most philosophers really think that truths of arithmetic are subject to empirical refutation together with any empirical theory in which they are embedded?) and we are surprised that a philosopher should think that a head count is a criterion of truth.

5. For canonical criticism of Quine on analyticity, see P. F. Strawson and H. P. Grice, "In Defense of a Dogma," *Philosophical Review* 1956. For more recent, meticulous criticism of Quine's general position, see H.-J. Glock, *Quine and Davidson on Language, Thought, and Reality* (Cambridge: Cambridge University Press, 2003). For the contrasts between Quine and Wittgenstein, see P. M. S. Hacker, *Wittgenstein's Place in Twentieth-Century Analytic Philosophy* (Oxford: Blackwell, 1996), chapter 7.

6. It might be thought (as suggested by Professor Churchland) that Descartes' view that the mind can causally affect the movement of the body (understood, according to Professor Churchland, as a conceptual claim) is refuted by the law of conservation of momentum. This is a mistake. It could be refuted (no matter whether it is a conceptual or empirical claim) only if it made sense; but, in the absence of criteria of identity for immaterial substances, it does not. The very idea that the mind is a *substance* of any kind is not coherent. Hence the statement that the mind, thus understood, possesses causal powers is not intelligible, a fortiori neither confirmable nor refutable by experimental observation and testing. (Reflect on what experimental result *would* count as showing that it is true.)

7. Such an *epistemic* conception informs Professor Timothy Williamson's lengthy attack on the very idea of a conceptual truth, "Conceptual Truth," *Proceedings of the Aristotelian Society,* suppl. vol. 80 (2006). The conception he outlines is not what many great thinkers, from Kant to the present day, meant by "a conceptual truth." Having criticized, to his own satisfaction, the epistemic conception that he himself delineated, Professor Williamson draws the conclusion *that there are no conceptual truths at all.* But that is a non sequitur of numbing proportions. For all he has shown (at best) is that there are no conceptual truths that fit the Procrustean epistemic bed he has devised.

8. The Aristotelian, anti-Cartesian, points that we emphasize are 1. Aristotle's principle, which we discuss below, 2. Aristotle's identification of the *psuchē* with a range of capacities, 3. that capacities are identified by what they are capacities to do, 4. that whether a creature possesses a capacity is to be seen from its activities, 5. Aristotle's realization that whether the *psuchē* and the body are one thing or two is an incoherent question.

9. It is, of course, not strictly a fallacy, but it leads to fallacies—invalid inferences and mistaken arguments.

10. A. J. P. Kenny, "The Homunculus Fallacy," in M. Grene, ed., *Interpretations of Life and Mind* (London: Routledge, 1971). We preferred the less picturesque but descriptively more accurate name "mereological fallacy" (and, correlatively, "the mereological principle"). We found that neuroscientists were prone to dismiss as childish the fallacy of supposing that there is a homunculus in the brain and to proceed in the next breath to ascribe psychological attributes to the brain.

11. Not, of course, *with* his brain, in the sense in which one does things *with* one's hands or eyes, nor in the sense in which one does things with one's talents. To be sure, he would not be able to do any of these things but for the normal functioning of his brain.

12. D. Dennett, *Content and Consciousness* (London: Routledge and Kegan Paul, 1969), p. 91.

13. We were more than a little surprised to find Professor Dennett declaring that his "main points of disagreement" are that he does not believe that "the personal level of explanation is the *only* level of explanation when the subject matter is human minds and actions" and that he believes that the task of relating these two levels of explanation is "not outside the philosopher's province" (p. 79). There is no disagreement at all over *this*. Anyone who has ever taken an aspirin to alleviate a headache, or imbibed excessive alcohol to become jocose, bellicose, or morose, and wants an explanation of the sequence of events must surely share Dennett's first commitment. Anyone who has concerned himself, as we have done throughout the 452 pages of *Philosophical Foundations of Neuroscience*, with clarifying the logical relationships between psychological and neuroscientific concepts, and between the phenomena they signify, share his second one.

14. L. Wittgenstein, *Philosophical Investigations* (Oxford: Blackwell, 1953), §281.

15. The Cartesian conception of the body a human being has is quite mistaken. Descartes conceived of his body as an insensate machine—a material substance without sensation. But our actual conception of our body ascribes verbs of sensation to the body we have—it is our body that aches all over or that itches intolerably.

16. The human brain is part of the human being. It can also be said to be part of the body a human being is said to have. It is, however, striking that one would, we suspect, hesitate to say of a living person, as opposed to a corpse, that his body *has* two legs or, of an amputee, that her body *has* only one leg. The misleading possessive is applied to the human being and to a human corpse, but not, or only hesitantly, to the body the living human being is said to have. Although the brain is a part of the human body, we surely would not say "my body *has* a brain" or "My body's brain has meningitis." That is no coincidence.

17. We agree with Professor Searle that the question of which of the lower animals are conscious cannot be settled by "linguistic analysis" (p. 104). But, whereas he supposes that it can be settled by investigating their nervous system, we suggest that it can be settled by investigating the behavior the animal displays in the circumstances of its life. Just as we find out whether an animal can see by reference to its responsiveness to visibilia, so too we find out whether an animal is capable of consciousness by investigating its behavioral repertoire and responsiveness to its environment. (That does not imply that being conscious is behaving in a certain way, but only that the criteria for being conscious are behavioral.)

18. The warrant for applying psychological predicates to others consists of evidential grounds. These may be inductive or constitutive (criterial). Inductive grounds, in these cases, presupposes noninductive, criterial grounds. The criteria for the application of a psychological predicate consist of behavior (not mere bodily movements) in appropriate circumstances. The criteria are defeasible. That such-and-such grounds warrant the ascription of a psychological predicate to another is partly constitutive of the meaning of the predicate but does not exhaust its meaning. Criteria for

the application of such a predicate are distinct from its truth-con-ditions—an animal may be in pain and not show it or exhibit pain behavior without being in pain. (We are no behaviorists.) The truth-conditions of a proposition ascribing a psychological predicate to a being are distinct from its truth. Both the criteria and the truth-conditions are distinct from the general conditions under which the activities of applying *or* of denying the predicate of creatures can significantly be engaged in. But it is wrong to suppose that a condition of "the language-game's being played" (as Professor Searle puts it) is the *occurrence* of publicly observable behavior. For the language game with a psychological predicate is played with its denial no less than with its affirmation. It would also be wrong to conflate the conditions for learning a language game with those for playing it.

19. J. Z. Young, *Programs of the Brain* (Oxford: Oxford University Press, 1978), p. 192. Professor Dennett also suggests (p. 90) that we mis-represented Crick in holding that, because he wrote that our brain believes things and makes interpretations on the basis of its previ-ous experience or information (F. Crick, *The Astonishing Hypoth-esis* [London:Touchstone, 1995], pp. 28–33, 57), therefore Crick really thought that the brain believes things and makes interpreta-tions, etc. We invite readers to look for themselves at Crick's cited discussions.

20. N. Chomsky, *Rules and Representations* (Oxford: Blackwell, 1980). Far from being oblivious to this, as Professor Dennett asserted (p. 91), the matter was critically discussed in G. P. Baker and P. M. S. Hacker, *Language, Sense and Nonsense* (Oxford: Blackwell, 1984), pp. 340–45.

21. D. Dennett, *Consciousness Explained* (Harmondsworth: Penguin, 1993), pp. 142–44.

22. Dennett here quotes from his autobiographical entry in S. Gutten-plan, ed., *A Companion to the Philosophy of Mind* (Oxford: Blackwell, 1994), p. 240.

23. Of course, we are not denying that analogical extension of concepts and conceptual structures is often fruitful in science. The hydrody-namical analogy generated a fruitful, testable, and mathematicized theory of electricity. Nothing comparable to this is evident in the

poetic license of Dennett's intentional stance. It is evident that poetic license allows Professor Dennett to describe thermostats as *sort
of believing* that it is getting too hot and so switching off the central
heating. But this adds nothing to engineering science or to the
explanation of homeostatic mechanisms.

Professor Dennett asserts (p. 88) that we did not address his attempts to use what he calls "the intentional stance" in explaining cortical processes. In fact we discussed his idea of the intentional stance at
some length (PFN 427–31), giving seven reasons for doubting its intelligibility. Since Professor Dennett has not replied to these objections,
we have, for the moment, nothing further to add on the matter.

In the debate at the APA Professor Dennett proclaimed that
there are "hundreds, maybe thousands, of experiments" to show
that a part of the brain has information that it contributes to "an
ongoing interpretation process in another part of the brain." This,
he insisted, is a "sort of asserting—a sort of telling 'Yes, there is col-
or here,' 'Yes, there is motion here.'" This, he said, "is just obvious."
But the fact that cells in the visual striate cortex fire in response to
impulses transmitted from the retina does not mean that they have
information or *sort of information* about objects in the visual field and
the fact that they respond to impulses does not mean that they *interpret* or *sort of interpret* anything. Or should we also argue that an
infarct shows that the heart has sort of information about the lack
of oxygen in the bloodstream and sort of interprets this as a sign
of coronary obstruction? Or that my failing torch has information
about the amount of electric current reaching its bulb and interprets this as a sign of the depletion of its batteries?

24. Thinking does not occur *in* the human being, but rather is done *by*
the human being. The *event* of my thinking that you were going to
V is located wherever I was located when I thought this; the *event*
of my seeing you V-ing is located wherever I was when I saw you V.
That is the only sense in which thinking, perceiving, etc. have a location. To ask, as Professor Searle does (p 110), where *exactly* did the
thinking occur, in any *other* sense, is like asking where exactly did a
person weigh 160 pounds in some other sense than that specified
by the answer "When he was in New York last year." Sensations,
by contrast, have a somatic location—if my leg hurts, then I have a

pain in my leg. To be sure, my state (if state it be) of having a pain in my leg obtained wherever I was when my leg hurt.

25. One needs a normally functioning brain to think or to walk, but one does not walk *with* one's brain. Nor does one think *with* it, any more than one hears or sees with it.

26. Professor Searle contends that because we repudiate qualia as understood by philosophers, therefore we can give no answer to the question of what *going through a mental process* consists in (p. 111). If *reciting the alphabet in one's imagination* (Professor Searle's example) counts as a mental process, it consists in first saying to oneself "a," then "b," then "c," etc. until one reaches "x, y, z." That mental process is not identified by its qualitative feel but by its being the recitation of the alphabet. The criteria for its occurrence include the subject's say-so. Of course, it can be supposed to be accompanied by as yet unknown neural processes, the locus of which can be roughly identified by inductive correlation using fMRI.

27. Descartes, *Principles of Philosophy* 1:46, 67, and especially 4:196.

28. As is asserted by Professor Churchland, "Cleansing Science," 469f., 474.

29. Ibid., p. 470.

30. For further discussion, see P. M. S. Hacker, *Wittgenstein: Meaning and Mind*, part 1: *The Essays* (Blackwell, Oxford, 1993), "Men, Minds and Machines," pp. 72–81.

31. C. Blakemore, "Understanding Images in the Brain," in H. Barlow, C. Blakemore, and M. Weston-Smith, eds., *Images and Understanding* (Cambridge: Cambridge University Press, 1990), p. 265.

32. J. Z. Young, *Programs of the Brain* (Oxford: Oxford University Press, 1978), p. 112.

33. D. Chalmers, *The Conscious Mind* (Oxford: Oxford University Press, 1996), p. 4.

34. F. Crick, *The Astonishing Hypothesis* (London: Touchstone, 1995), pp. 9f.

35. A. Damasio, *The Feeling of What Happens* (London: Heineman, 1999), p. 9.

36. Ned Block, "Qualia," in S. Guttenplan, ed., *Blackwell Companion to the Philosophy of Mind* (Oxford: Blackwell, 1994), p. 514.

37. Searle, *Mystery of Consciousness* (London: Granta, 1997), p. xiv.

38. T. Nagel, "What It Is Like to Be a Bat?" reprinted in his *Mortal Questions* (Cambridge: Cambridge University Press, 1979), p. 170.

39. Professor Searle (like Grice and Strawson) supposes that perceptual experiences are to be characterized in terms of their highest common factor with illusory and hallucinatory experiences. So all perceptual experience is, as it were, hallucination, but veridical perception is a hallucination with a special kind of cause. This, we think, is mistaken.

40. Professor Searle asserts that we deny the existence of qualitative experiences (p. 99). We certainly do not deny that people have visual experiences, i.e., that they see things. Nor do we deny that seeing things may have certain qualities. What we deny is that whenever someone sees something, there is something it is like for them to see that thing, let alone that there is something it feels like for them to see what they see. And we deny that "the *qualitative* feel of the experience" is its "defining essence" (p. 115). Seeing or hearing are not defined by reference to what they feel like, but by reference to what they enable us to detect.

41. G. Wolford, M. B. Miller, and M. Gazzaniga, "The Left Hemisphere's Role in Hypothesis Formation," *Journal of Neuroscience* 20 (2000), RC 64 (1–4), p. 2.

42. We are grateful to Robert Arrington, Hanoch Ben-Yami, Hanjo Glock, John Hyman, Anthony Kenny, Hans Oberdiek, Herman Philipse, Bede Rundle, and especially to David Wiggins for their helpful comments on the early draft of this paper, which we presented at the APA, Eastern Division, in New York on December 28, 2005, in an "Authors and Critics" debate.

EPILOGUE

1. Professor Searle suggested this at the APA meeting. He claimed that the attainment of this goal will proceed in three steps: first, determination of the neural correlates of consciousness (NCC); second, establishment of the causal relationship between consciousness and these NCC; and, finally, development of a general theory relating consciousness and the NCC.

2. J. Searle, *Mind: A Brief Introduction* (Oxford: Oxford University Press, 2004), chapter 5.

3. J. Searle, "Consciousness: What We Still Don't Know," *New York Review of Books*, January 13, 2005. This is a review of Christof Koch, *The Quest for Consciousness* (Greenwood Village, CO: Roberts, 2004).

STILL LOOKING

1. A most discerning essay on this work is Lana Cable, "Such nothing is terrestriall: philosophy of mind on Phineas Fletcher's Purple Island." *Journal of Historical Behavioral Science* 19(2): 136–52.

2. Aristotle, *On the Soul*, 403a25–403b1, in Richard McKeon, ed., *The Basic Works of Aristotle*, J.A. Smith, trans. (New York: Random House, 1941).

3. Ibid., 408b10–15.

4. This is in his *Philosophical Investigations*, §265.

CPSIA information can be obtained
at www.ICGtesting.com
Printed in the USA
JSHW041544180121
11014JS00003B/62